Source Water Protection

Operational Guide to ANSI/AWWA G300

Richard W. Gullick, PhD

2016

Operational Guide to ANSI/AWWA G300
Source Water Protection

Project Manager: Melissa Valentine
Production: PerfecType, Nashville, TN

Printed in the United States of America
American Water Works Association
6666 West Quincy Ave.
Denver, CO 80235

Contents

American Water Works Association

G300 Operational Guide

Source Water Protection

SECTION 1: ACKNOWLEDGMENTS

This *Operational Guide to AWWA Standard G300: Source Water Protection* is the result of a series of ongoing efforts over the past decade and a half. In the fall of 2000, the American Water Works Association (AWWA) requested that the volunteer AWWA Source Water Protection (SWP) Committee develop accreditation criteria and supporting guidance for SWP activities performed by water utilities. Led by SWP Committee chair Pamela P. Kenel, P.E. (Black & Veatch), a workgroup of approximately 18 SWP professionals from across the United States met in Kansas City, Kansas, on December 4–5, 2000, to develop an outline of the accreditation criteria. The majority of the worksheets included in this guide were developed at that workshop. Members of the workgroup included the following:

Pamela P. Kenel (Chair), Black & Veatch Engineers, Gaithersburg, Md.

Christopher S. Crockett, Philadelphia Water Department, Philadelphia, Pa.

Andrew F. DeGraca, San Francisco Public Utilities Commission, San Francisco, Calif.

Richard W. Gullick, American Water Works Service Company, Voorhees, N.J.

Betsy Henry, US Environmental Protection Agency, Washington, D.C.

Rachael R. Herpel, Groundwater Foundation, Lincoln, Neb.

Edward A. Holland, Orange Water & Sewer Authority, Carrboro, N.C.

Ronald B. Hunsinger, East Bay Municipal Utility District, Oakland, Calif.

1

Gary W. Jackson, US Department of Agriculture/National Resources Conservation Service, Madison, Wis.

Florence Reynolds, Salt Lake City Department of Public Utilities, Salt Lake City, Utah

Daniel S. Schechter, LimnoTech, Ann Arbor, Mich.

Perri Standish-Lee, Black & Veatch Engineers, Granite Bay, Calif.

Ira A. Stern, New York City Department of Environmental Protection, Corona, N.Y.

Kenneth A. Thompson, Irvine Ranch Water District, Irvine, Calif.

John T. Witherspoon, Springfield City Utilities, Springfield, Mo.

Clare Haas Claveau, AWWA (liaison), Denver, Colo.

William C. Lauer, AWWA (liaison), Denver, Colo.

Steve Via, AWWA (liaison), Washington, D.C.

A draft version of the Source Water Protection and Management Program Accreditation document was completed in June 2002, led by Dr. Richard W. Gullick (American Water).

AWWA's planned accreditation program evolved into a new series of utility management standards for water and wastewater utilities (described in the Foreword). The AWWA Standards Committee on Source Water Protection was formed and chaired by Edward A. Holland (Orange Water & Sewer Authority). That committee approved ANSI/AWWA G300, Standard for Source Water Protection, in 2006, and the AWWA Board of Directors approved the standard in 2007.

AWWA then requested the SWP Committee to edit the 2002 draft accreditation document into an Operational Guide to ANSI/AWWA Standard G300. The text was expanded, a few unfinished worksheets were completed, and three case studies and more extensive lists of resources were added. Coauthors of the 2010 version of the guide included Dr. Chi Ho Sham (Cadmus Group Inc.), Dr. Richard W. Gullick (Environmental Engineering & Technology, Inc.), Dr. Sharon C. Long (Wisconsin State Laboratory of Hygiene and University of Wisconsin–Madison), and Pamela P. Kenel, P.E. (Black & Veatch); all four are former chairs of the AWWA SWP Committee.

The AWWA Standards Committee on Source Water Protection (chaired by Dr. Chi Ho Sham, Cadmus Group Inc.) updated ANSI/AWWA Standard G300 in 2014. This revised version of the Operational Guide to ANSI/AWWA Standard G300 was prepared in 2016 by Dr. Richard W. Gullick (Rivanna Water & Sewer Authority).

AWWA is grateful to the following drinking water utilities that allowed their SWP programs to be described as case studies in this guide: Philadelphia Water (Pa.), Central Arkansas Water (Little Rock), and Remsen Municipal Utilities (Iowa). The following personnel contributed to the writing of their respective case study descriptions:

> Kelly Anderson, Source Water Protection Program Manager, Philadelphia Water
>
> Elizabeth Ventura, Source Water Protection Program, Philadelphia Water
>
> Alison Aminto, Source Water Protection Program, Philadelphia Water
>
> Molly D. Hesson, Source Water Protection Program, Sage Services LLC (for Philadelphia Water)
>
> Paul R. "Randy" Easley, Director of Water Quality, Central Arkansas Water
>
> John Tynan, Watershed Protection Manager, Central Arkansas Water
>
> Becky Ohrtman, Source Water Protection Coordinator, Iowa Department of Natural Resources (for Remsen)
>
> Dr. Chi Ho Sham, Chief Scientist and Vice President, Eastern Research Group (for Remsen)

The following subject experts reviewed this guide and provided constructive commentary: Nancy Toth, Eugene Water & Electric Board; Paul R. "Randy" Easley, Director of Water Quality, Central Arkansas Water; Pamela P. Kenel, P.E., Black & Veatch; and Jonathan Keck, California Water Service Company. AWWA staff members Paul Olson, Bill Lauer, Dawn Flancher, John Anderson, Martha Ripley Gray, Daniel Feldman, and Gay Porter DeNileon also contributed to this effort.

SECTION 2: FOREWORD

AWWA's primary goal is to support water utilities in the evaluation and improvement of their water quality, operations, maintenance, and infrastructure. Several programs and types of publications are used to support this mission.

A key program is the AWWA Standards Program, which has existed for more than 100 years to produce peer-reviewed standards for the materials and processes used by water and wastewater utility industries. These standards, which are American National Standards Institute (ANSI) approved, are recognized worldwide and have been adopted by many utilities and organizations. The AWWA Standards Program is designed to assist water and wastewater utilities and their service providers in meeting the expectations of their customers, investors, and government regulators. The standards developed under the program are generally intended to improve a utility's overall operations and service.

Recently, AWWA also developed a series of management standards for water and wastewater utilities. The Utility Management Standards Program provides a means to assess service quality and management efficiency based on recognized standards for best available practices. Through these standards and formal recognition by professional organizations, the program serves water and wastewater utilities by promoting improvements in the quality of services and efficient management.

The utility management standards address the utility manager's need to have consistency and reliability and to know what is expected in the management and operation of a utility. These standards are also valuable resources for the many issues that utilities face, including increased scrutiny on accountability, increased regulation, and difficult economic realities such as aging infrastructure, changing demand for water, and a shrinking workforce.

The utility management standards are designed to cover the principal activities of a typical water and/or wastewater utility and include the following:

- G100, Water Treatment Plant Operation and Management
- G200, Distribution Systems Operation and Management
- G300, Source Water Protection
- G400, Utility Management System
- G410, Business Practices for Operation and Management
- G420, Communications and Customer Relations
- G430, Security Practices for Operation and Management
- G440, Emergency Preparedness Practices
- G480, Water Conservation Program Operation and Management
- G481, Reclaimed Water Program Operation and Management
- G510, Standard for Wastewater Treatment Plant Operations and Management

The utility management standards are developed using the same formal, ANSI-recognized, AWWA-managed process. Volunteers on the standards committees establish standard practices in a uniform and appropriate format. Formal standards committees are formed to address the individual standards of practice for the diverse areas of water and wastewater utility operations.

ANSI/AWWA Standard G300, Source Water Protection, is the definitive standard for a drinking water utility to use to protect its drinking water supply source(s). The first edition of ANSI/AWWA G300 became effective on July 1,

2007, and was revised on June 1, 2014. The standard outlines the six primary components of successful source water protection (SWP) programs and the requirements for meeting the standard.

Given that every water system is unique, SWP elements vary from utility to utility. The standard can be used to guide development of a SWP program by following the framework presented. Also, it can be used to evaluate the completeness and potential effectiveness of an existing SWP program by comparing a program's components with those listed in the standard to ensure all important parts are included.

This guide is intended to provide support and guidance to help utilities plan, develop, and implement successful SWP programs that will meet the criteria listed in ANSI/AWWA G300. However, it is not intended to be a complete instructional guide for implementation of ANSI/AWWA G300 and may not apply to all situations encountered by individual utilities. It also does not have any specific regulatory linkage and does not specifically address source water selection or certain aspects of source water management (e.g., treatment of reservoir water to control algal growth and stratification).

This guide starts with acknowledgments (Sec. 1) of key individuals who assisted in the development of ANSI/AWWA G300 and this associated operational guide. Section 2 (this Foreword) describes the basic tenets of the AWWA Utility Management Standard Program and describes the outline of this guide.

Section 3, Introduction, includes basic information and guidance about SWP programs and activities and a brief description of the six essential elements (or phases) that comprise a successful SWP program. This section explains how this guide can be used to assess whether a SWP program has met the criteria of ANSI/AWWA Standard G300.

Section 4, Requirements, lists the specific requirements of the SWP standard. There are six subsections, one for each of the six basic elements of a successful SWP program. Each subsection includes a description of the rationale for the specific requirement, a description of the program component and related issues, some suggested approaches, and, for some elements, a list of resources for further information.

Section 5, Verification, includes a description of documentation and human resources needed to verify compliance with the standard.

Section 6, Glossary of Abbreviations, lists the abbreviations used in this guide.

Section 7, Resources, lists relevant national-level stakeholders.

Section 8, Annotated Bibliography for Select Information Sources, provides descriptions and web links for helpful reports, websites, databases, and other sources of information and guidance.

Section 9, References, lists all references cited in this guide.

Section 10, Worksheets, contains many questions in checklist format that utilities can use to self-assess their progress. These questions serve as the basis for reporting progress related to ANSI/AWWA G300, Standard for Source Water Protection. These worksheets are not intended to serve as blueprints to all SWP programs. However, a utility that has all the components represented in the checklists is likely to have the principle elements of the ANSI/AWWA standard incorporated into its SWP program and practices. The worksheets can be used to ensure that important subjects are covered for each component, to find additional ideas that may not have been included in the existing program, and to help identify gaps that may exist in a utility's current procedures compared to the requirements of the standard. The worksheets also serve to solicit proof and documentation that procedures are indeed in place where called for by the standard.

Section 11, Case Studies for Source Water Protection, describes three successful SWP programs for the following utilities: Philadelphia Water (Pa.), Central Arkansas Water (Little Rock), and Remsen Municipal Utilities (Iowa). Each case study follows the framework of the ANSI/AWWA Standard and thus provides examples of different ways to accomplish each necessary component. They also show innovative elements that go beyond the minimum requirements of the standard. Philadelphia Water and Remsen Municipal Utilities are past recipients of the AWWA Exemplary Source Water Protection Award, which is based on following the concepts laid out in the standard.

SECTION 3: INTRODUCTION

The goal of drinking water suppliers is to provide their customers with sufficient quantities of high-quality water at affordable rates. The best way to ensure high-quality drinking water for public consumption is to use a multiple barrier approach (e.g., O'Connor 2002; CCME 2002). This approach includes (1) selecting the highest-quality source water possible; (2) protecting that source; (3) treating the water (with more than one method); (4) maintaining quality in the distribution system; (5) monitoring quality at those various stages; and (6) when necessary (if other barriers fail), implementing adequate emergency response procedures. Sound management practices and an adequate regulatory structure provide a method and a framework for proper implementation of this approach. Each step in this process should be optimized when responding to high customer expectations for drinking water quality, increased regulatory requirements, and uncertainties associated with the growing number of drinking water contaminants as well as climate change.

Given the great importance of source water protection (SWP) as one component of the multiple-barrier approach, ANSI/AWWA G300, Source Water Protection, was developed. First published in 2007, it was revised in 2014 (AWWA 2014). The standard was established to provide a general framework to help guide drinking water systems of any size and to provide the conditions for developing SWP programs that are locally specific and highly variable in scope. It also provides a tool for assessing the completeness and effectiveness of a SWP program.

Background

Contaminants of concern for drinking water result from either natural sources or anthropogenic activities within the water supply watersheds and aquifers and from unintended collateral effects of the treatment processes used to deal with certain contaminants in source waters (e.g., disinfection by-products). Therefore, high-quality source water is an important barrier in preventing contaminants from entering or being created within the water supply system. AWWA's support of this premise is noted in its official policy statement on the quality of water supply sources: "AWWA strongly supports securing drinking water from the highest quality sources available and protecting those sources to the maximum degree possible" (AWWA 2010).

The primary objectives of SWP programs are to maintain, safeguard, and/or improve the quality of a given water source. A clear and important aspect of this premise is that pollution prevention is often preferable to remediation or treatment of contaminated source water. SWP programs should not only emphasize the reduction in current contamination but also provide a means to assess and prevent future contamination.

Numerous benefits may be achieved from SWP, and each may be considered an incentive or driver for a utility to develop and implement SWP activities. The potential incentives include the following:

- Increasing public health protection by ensuring higher-quality raw water (and thus presumably higher-quality treated water), especially for sensitive subpopulations. In addition to reduction in illnesses and mortalities, public health protection can provide economic benefits in terms of reduced health care costs and reduced loss of productivity and work time.
- Supporting water and land resource sustainability goals, legislation, and regulation.
- Increasing the likelihood of complying with existing and future drinking water regulations (including maximum contaminant levels).
- Qualifying for a 0.5-log removal credit for watershed control programs under the US Environmental Protection Agency's (USEPA's) Long-Term 2 Enhanced Surface Water Treatment Rule.
- Providing a means to respond to uncertainties presented by the growing number of unknown or unregulated microbiological and chemical contaminants (i.e., preventing contamination that treatment may not remove).
- Avoiding costs of contamination, including
  - Reducing water treatment challenges and costs (e.g., more consistent influent water quality, and lower current and future capital and operating treatment costs)
  - Saving potential future expenses associated with land and water contamination remediation
  - Saving monitoring, engineering, and legal expenses
  - Saving expenses related to finding and obtaining alternate water supplies
  - Reducing the potential for more indirect financial costs, such as real estate devaluation or lost jobs and tax revenue
- Maintaining or improving source water quality for uses other than drinking (e.g., fish habitat, recreation)
- Protecting aesthetic water quality (i.e., prevention of taste, odor, and color problems)
- Meeting utility customer expectations, and improving or preventing a decline in customer/citizen perceptions and confidence
- Providing for general environmental stewardship for current and future generations (e.g., improving the overall environmental quality of watersheds and aquifers)
- Receiving recognition as a steward of the environment

- Maintaining or improving bond ratings, for example, by improving the overall resiliency of the water system
- Increasing funding opportunities (e.g., in some cases, communities with SWP programs may receive higher priority for low-cost loans and grants related to drinking water or watershed management programs)
- Increasing aesthetic beauty and/or economic value of residential and commercial properties through use of attractive best management practices (BMPs) such as artificial ponds or wetlands
- Improving communication and cooperation among stakeholders. Effective collaboration can at times help drinking water utilities avoid the need to play the primary role and allow them to be a supporting catalyst for change and activity
- Improving operations and reducing expenses for various industries and commercial establishments (e.g., nutrient management plans may reduce the need for fertilizer, sediment control BMPs can maintain soil resources, and improvement in energy and water efficiency can reduce carbon and water footprints)
- Realizing social benefits and greatly improved relations with upstream neighbors resulting from SWP efforts

To minimize impacts from chemical and microbiological contaminants, pollutant sources that SWP programs can address include the following:

- Agricultural, commercial, and residential use of fertilizers and pesticides
- Fuel and other chemical use, storage, transportation, and disposal (e.g., aboveground and underground storage tanks)
- Accidental spills or releases of contaminants (e.g., transportation accidents)
- Deliberate spills or releases of contaminants (e.g., vandalism and terrorism)
- Mining and oil and gas extraction
- Solid waste and hazardous waste disposal sites
- Commercial and industrial establishments
- Changes in land-use patterns such as new residential and commercial development or intensification of agriculture
- Stormwater runoff from natural events that involve contaminants such as microorganisms, nutrients, heavy metals, organic chemicals, and sediment
- Treated and untreated municipal and industrial wastewater discharges
- Combined and sanitary sewer overflows
- Septic systems

- Abandoned, injection, and production wells
- Animal waste from livestock, pets, and wildlife
- Highways and transportation systems such as airports (including deicing and other maintenance activities) and railroads

Lists of various regulated and unregulated drinking water contaminants can be found on the USEPA website (www.epa.gov/safewater/contaminants/index. html). These lists include contaminants that are subject to the National Primary Drinking Water Regulations and National Secondary Drinking Water Regulations. USEPA lists of unregulated contaminants include those found on the Drinking Water Contaminant Candidate List website (www.epa.gov/safewater/ccl/index. html), those subject to the Unregulated Contaminant Monitoring Program (www. epa.gov/dwucmr), and those included in the National Contaminant Occurrence Database   (www.epa.gov/dwucmr/national-contaminant-occurrence-database-ncod). Additional chemicals of interest can be found through the USEPA's Toxic Release Inventory Program (www.epa.gov/toxics-release-inventory-tri-program).

A wide variety of technical and managerial SWP practices are available for use, including the following:

- Contaminant source reduction and management
- Management practices for point and nonpoint pollution sources
- Stormwater management
- Wastewater treatment plant upgrades and maintenance
- Rules and assistance for maintaining septic systems
- Agricultural management practices, incentives, and land stewardship programs
- Erosion and sediment control for construction projects
- Regulations and permits such as construction and operating standards, equipment operation and maintenance, and public health regulations
- Land-use controls, including zoning, subdivision growth controls, acquisition of development rights, conservation easements, land-use prohibitions, and land purchases
- Forestry management programs
- Wildlife control
- Visual inspection of water supplies and potential contaminant sources
- Source water monitoring (including early warning monitoring for spill events and chemical and microbial pollutant source tracking)
- Spill prevention, control, and countermeasure plans and implementation

- Emergency response (contingency) planning and implementation
- Stakeholder coalitions and collaborations
- Sanitary surveys
- Education programs

Numerous resources are available for detailed guidance; some are listed in Secs. 7, 8, and 11. These include guides for developing and implementing SWP programs, information on monitoring methods, lists of funding possibilities, manuals and databases that describe implementation and effectiveness of management practices, and other related topics.

Basic Elements of the Source Water Protection Standard

SWP is a highly site-specific process. Different water sources require widely different approaches. For example, substantially different SWP programs would be appropriate for pristine mountain streams, for the lower reaches of highly developed rivers, and for groundwater supplies. Even similar types of water supplies may require different program components as a result of the differing characteristics of their watersheds, accompanying land uses, potential contaminant sources, active stakeholders, and available resources. Therefore, a general framework for the development and implementation of SWP programs, as opposed to contaminant-specific guidance and ratings, is the most appropriate approach for a universal standard. Within this broad framework, specific SWP programs must account for local conditions, incorporate diverse stakeholder interests, seek commitment to the SWP process by all involved parties, and strive to be sustainable over the long term.

Six main elements (or steps) comprise the process of developing and implementing a successful SWP program at the water utility level, as shown in Figure 3-1, and ANSI/AWWA G300, Standard for Source Water Protection, is based on these steps. The primary objectives and components of each element are discussed in the following sections:

- A SWP program *vision* and *stakeholder involvement*
- Source water *characterization*
- SWP *goals*
- SWP *action plan*
- *Implementation* of the action plan
- Periodic *evaluation and revision* of the entire program

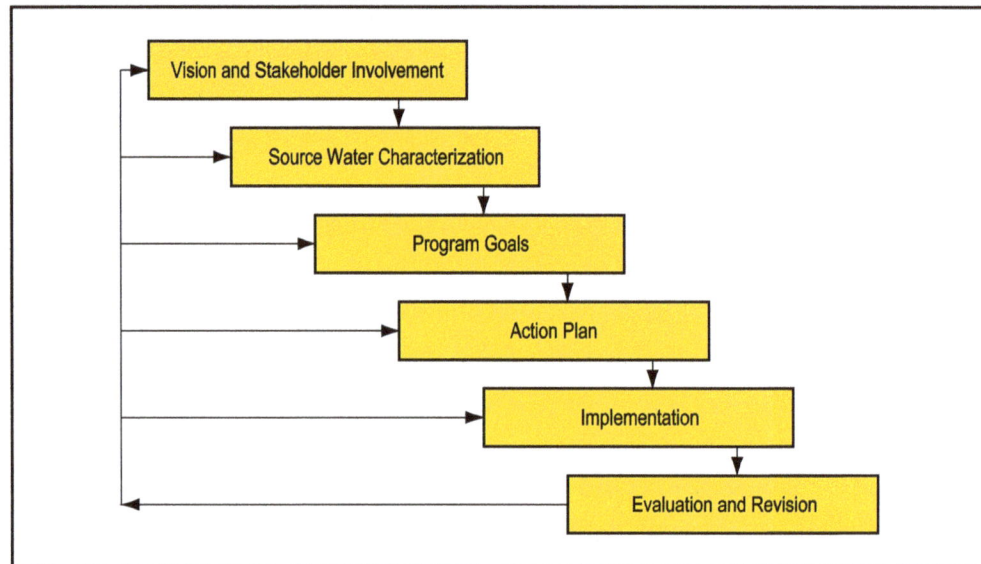

Figure 3-1 Six Elements of Successful Source Water Protection Programs

Source: Modified from Gullick (2003) and AWWA (2014).

Although each of the six primary steps may differ greatly in terms of complexity or effort, each step is vital to the success of every SWP program. Accordingly, basic success in each area must be demonstrated in order for a utility to meet the criteria of ANSI/AWWA G300 Standard for Source Water Protection (AWWA 2014).

Source Water Protection Vision

A formalized vision guides the development and implementation of a SWP program. The vision helps to align priorities and resources for the effort. Involvement of outside stakeholders is usually essential for a successful SWP program. It is best to involve stakeholders throughout the process, including in development of the vision.

Source Water Characterization

Characterization and assessment of the source water and the land or subsurface area from which it is derived provide information for conducting a source water susceptibility analysis that is used to identify and prioritize the key water quality issues and contamination sources. Program goals are then developed based on this analysis and the utility's and other stakeholders' priorities and capabilities.

Program Goals

Goals and objectives need to be formulated to guide the SWP program and its specific elements. The goals should be developed in response to specific problem areas identified through the source water characterization and risk assessment processes, and they should address each driver that motivates the SWP program, including the SWP vision. Goals, which may address both current and potential future issues, should be prioritized to reflect the concerns of greatest importance and ideally should specify temporal and qualitative and/or quantitative dimensions (e.g., specific timelines and measurable goals). Both internal and external stakeholders should be involved in development of the goals.

Action Plan

The action plan lays out a road map of activities to conduct in order to achieve the desired watershed protection goals based on the vision, source water area characterization, and susceptibility analysis. The plan identifies actions that are required (e.g., regulations, agreements, practices) to mitigate existing and future threats to source water quality, develops priorities for implementation, and includes a timetable for implementation, identification of necessary resources and ways to obtain those resources (e.g., funding), and metrics for measuring success of each plan component. These priorities may be based on the perceived risk from different contaminant sources, the resources available to implement actions, the likelihood of success of different actions, and the obstacles to success that exist for different contaminant sources.

Implementation of the Action Plan

Implementation of the action plan is the core of any SWP program. Planning without implementation does not provide results. Without this step, no actual protection takes place. Development of a comprehensive and implementable plan supports effective implementation, and use of an adaptive and iterative management approach can help to respond to unexpected challenges and barriers.

Source Water Protection Program Evaluation and Revision

Any type of administrative program requires periodic (or continuous) evaluation and revision. A good SWP program includes provisions for reviewing and, if necessary, modifying the utility's SWP vision, characterization, goals, action plan, and implementation elements. This should be done periodically and also in response to changes in the source water area, changes in contaminant sources, and performance of implemented programs. This step is intended to measure the accomplishment or completion of projects, programs, and activities identified in the action plan and to identify obstacles to success and ways to overcome those obstacles. The evaluation and revision aspects of the SWP program should be designed as iterative and interactive processes. SWP plans should be living documents that are continuously improved and updated.

ANSI/AWWA G300 Assessment Process

As noted previously, each of the six primary steps in the SWP program framework are likely to differ greatly in terms of complexity or effort, but each step is vital to the success of any SWP program. Accordingly, basic success in each area must be demonstrated in order for a utility to meet the criteria of ANSI/AWWA G300, Standard for Source Water Protection.

To provide guidance and serve as a tool to assist in the assessment process, 23 worksheets with a series of questions to be considered are provided in Sec. 10. There is one worksheet for each element of the SWP program process (worksheets A, B, C, D, E, and F). Additional worksheets with greater detail are provided for the characterization phase (worksheets B-1 through B-8), planning (worksheets D-1 and D-2), and program implementation (worksheets E-1 through E-6). A final worksheet (worksheet G) covers the verification and record-keeping components of the standard. General qualitative assessment criteria are provided for the answers to most questions via a high/medium/low (or high/medium/low/none) rating system, with "high" being the best rating.

Note that the answers to specific questions may be qualitative or subjective, and specific quantitative criteria are often difficult to develop given the inherent diversity between different source water land areas, different water suppliers, and other factors. In addition to following the six steps in the SWP process, the evaluation should also consider how well the utility has tailored its program to fit the specific needs of its water sources and the utility's unique characteristics.

Note that SWP is a process that typically spans more than one year and a variety of external obstacles from outside of the water utility could be encountered (e.g., Sklenar et al. 2012; Gullick 2014). Therefore, although it would be ideal to complete or make progress on each primary step in SWP, it is not necessary for a utility to complete all aspects of the implementation and evaluation phases of its SWP program in order to meet the standard's requirements. A comprehensive SWP plan should be used to guide the implementation and evaluation of SWP activities. Nevertheless, a well-conceived plan without on-the-ground execution of specific activities would be just a paper exercise that does not offer any real public health protection or improvement to a utility's overall operations and service.

SECTION 4: REQUIREMENTS

This section discusses the six main steps (Figure 3-1) in the process of developing and implementing a successful source water protection (SWP) program at the water utility level. These steps comprise the basis of ANSI/AWWA G300, Standard for Source Water Protection. Text in *italics* is quoted from the standard.

Sec 4.1   Source Water Protection Program Vision and Stakeholder involvement

A formalized vision guides the development and implementation of a source water protection program. A vision statement explicitly expresses the policy of the organization as set forth by the decision-making body of the utility and helps to align priorities and resources. The vision or policy statement is the official declaration of the utility's commitment to source water protection.

4.1.1   Vision

The utility shall have a vision or policy that expresses a commitment to source water protection. The vision or policy (or other similar utility document, or consistent practice) shall include a commitment of, or intention to commit, sufficient resources to the source water protection effort. A written version of the vision is strongly encouraged because it would serve to preserve institutional history and allow future generations to be reminded of the vision, along with the opportunities to review and update the vision.

4.1.2   Stakeholder Involvement

Involvement of relevant outside stakeholders is usually essential for development and implementation of a successful source water protection program. The utility shall identify source water area stakeholders, their roles, and existing initiatives in which they may be engaged. Cooperation or partnerships with relevant stakeholders shall be realistically assessed and actively pursued throughout program development and implementation. The formation of collaborative associations at the watershed and local levels is an effective approach to engage stakeholders and partners. Stakeholder involvement may result in improved coordination of partnership activities, additional volunteer efforts, and potential funding opportunities. It is expected that various stakeholders may be involved in each stage of the source water protection program process, but also that not all of the same stakeholders may be involved in each of the stages.

Rationale

Having a vision will help in setting goals and in planning and implementing the program. The vision states a utility's SWP policy as put forth by the utility's decision-making body and helps to align priorities and resources. This commitment from the governing body is important as it helps to ensure staff have the necessary supporting resources to not only plan but to implement SWP practices.

The vision may be articulated in a vision/mission statement or policy. It is the utility's "call to action" and is a statement of the utility's commitment to SWP. As such, the AWWA standard requires that a utility have a vision or policy statement that formally expresses a commitment to SWP or at least a consistent practice of committing sufficient resources to the SWP effort.

A vision/mission statement may include any number of the following considerations:

- Protecting public health
- Protecting against taste and odor
- Reducing treatment costs
- Increasing reliability of treatment technology
- Increasing supply reliability
- Knowing what contaminants are regulated
- Addressing contaminants that are emerging or unknown
- Providing regulatory incentives
- Addressing needs of sensitive subpopulations
- Knowing customer expectations and perceptions
- Balancing risks
- Protecting investment in water sources
- Being aware of bond ratings
- Performing environmental stewardship and sustainability
- Knowing the objectives of other stakeholders

The long-term success of a SWP program often depends on the acceptance, buy-in, and participation of key stakeholders. These stakeholders can include other water users and suppliers; government officials; important commercial, industrial, and agricultural interests; environmental and citizen groups; landowners; and the public. Their contributions and buy-in to the vision can help to garner support for subsequent phases and can form the initial framework for planning and implementing the process. Accordingly, basic principles of good stakeholder involvement and communication are discussed here. These principles also apply to the other basic elements and phases of successful SWP programs.

The start of the process, when the vision is being developed, is an appropriate stage for collaborating with a variety of stakeholders who share a common interest in the water supply, land use, SWP, and other related issues. It is important that the program's vision, mission, and goals are clearly defined at the outset and that the all stakeholders have an opportunity to provide input and receive information at all stages. These stakeholders typically do not expect every decision or action to benefit them directly; however, clearly defining and adhering to a formal and transparent decision-making process usually helps to generate a level of comfort with the various stakeholders.

The purpose of a stakeholder involvement program is at least fourfold: (1) to bring diverse stakeholders together so everyone involved can understand the benefits and impacts of the proposed program, (2) to jointly identify a variety of opportunities and alternatives, (3) to bring various resources, such as expertise, staffing, and funding, to the program effort; and (4) to identify common ground that relates to the improvement of watershed health and/or aquifer conditions.

An initial step is to understand and describe each stakeholder's goals and concerns, as well as to identify watershed protection practices currently used or supported by each stakeholder and possible constraints that prevent implementation of other methods. It is important that stakeholders understand the mechanisms that will be used to allow input into the program and related products, such as reports and results of analyses. In this way, stakeholders will not view technical analyses as mysterious and will be able to meaningfully participate in the program.

Workshops with the stakeholders should be held at key decision points throughout the program. Experience shows that workshops ensure that effective communication takes place in an efficient manner and allow for everyone's concerns and ideas to be identified and addressed. An external facilitator who is not connected to any of the stakeholders in the program can implement the workshops. The facilitator should provide support before and after each workshop to ensure that the time allocated to these meetings is highly productive. Orientation material, including an agenda, appropriate reference materials, and a clear description of the desired outcomes, should be distributed to the participants before each workshop. Following each workshop, summary meeting notes should be prepared and disseminated.

Also, open meetings should be held at critical junctures of the program to inform the general public and obtain input. Other outreach techniques such as newsletters, flyers, and websites are also effective ways to inform the public.

A program website can enhance communications among stakeholders and other interested parties and is an efficient repository of information. Websites can be made secure from anyone without the appropriate passwords, allowing only authorized staff and stakeholders access to project information. If appropriate, a read-only component could be linked to the website to enhance access of project information to the general public.

Worksheets

See Worksheet A, Developing a Vision and Coordinating with Stakeholders, in Sec. 10 for guidance in developing a vision and working with stakeholders.

Resources

Fletcher, Angie, Susan Davis, and Grantley Pyke. 2005. *Water Utility/Agricultural Alliances: Working Together for Cleaner Water*. Denver, Colo.: Water Research Foundation.

Herpel, Rachael. 2004. *Source Water Assessment and Protection Workshop Guide. 2nd ed.* Lincoln, Neb.: Groundwater Foundation. http://www.groundwater.org/action/resources-info.html.

Raucher, Robert S., and James Goldstein. 2001. *Guidance to Utilities on Building Alliances with Watershed Stakeholders*. Denver, Colo.: Water Research Foundation.

Sklenar, Karen and Laura J. Blake. 2010. *Drinking Water Source Protection Through Effective Use of TMDL Processes*. Denver, Colo.: Water Research Foundation.

Sklenar, Karen, Chi Ho Sham, and Richard W. Gullick. 2012. *Source Water Protection Vision and Roadmap*. Denver, Colo.: Water Research Foundation.

Source Water Collaborative. 2016. Collaboration Toolkit (online guide). www.sourcewatercollaborative.org/how-to-collaborate-toolkit.

USEPA (US Environmental Protection Agency). 1999. *Protecting Sources of Drinking Water: Selected Case Studies in Watershed Management*. EPA 816-R-98-019, April 1999. Washington, D.C.: USEPA Office of Water.

———. 2000. *Watershed Success Stories—Applying the Principles and Spirit of the Clean Water Action Plan*. Washington, D.C.: USEPA.

————. 2001a. *National Source Water Contamination Prevention Strategy: Seventh Draft for Discussion* (April 2001). Washington, D.C.: USEPA Office of Ground Water and Drinking Water. http://permanent.access.gpo.gov/lps21800/www .epa.gov/safewater/protect/strateg7.pdf; A companion discussion of national SWP information needs is also available at http://permanent.access.gpo.gov /lps21800/www.epa.gov/safewater/protect/1205meas.pdf.

————. 2008. *Handbook for Developing Watershed Plans to Restore and Protect Our Waters.* EPA 841-B-08-002. Washington, D.C.: USEPA Office of Water. www .epa.gov/polluted-runoff-nonpoint-source-pollution/handbook-developing -watershed-plans-restore-and-protect.

————. 2010. *Getting in Step: A Guide for Conducting Watershed Outreach Campaigns, 3rd Edition.* EPA 841-B-10-002, 163+ pp. Washington, D.C.: USEPA. https://cfpub.epa.gov/npstbx/files/getnstepguide.pdf.

Sec. 4.2   Characterization of Source Water and Source Water Protection Area

[This is the initial information collection and analysis phase of SWP programs.] Characterization and assessment of the source water, the land or subsurface area from which the source water is derived, and the potential sources of contamination (PSCs) in those areas is essential for obtaining the understanding and knowledge needed to develop the goals and plans that will realize the source water protection vision. Using that information, a risk assessment or susceptibility analysis is conducted to identify and prioritize the key water quality and SWP issues and contamination sources. In some cases, it will be appropriate and expected that users of this standard will have gone beyond state-performed source water assessments to better define watershed characteristics and have obtained extensive public participation in defining implementation components of the source water protection program.

4.2.1   **Delineation**

The utility shall geographically delineate its source waters and the areas of concern from which they are derived. This standard is valid for both groundwater and surface water.

4.2.2   **Water Quality and Quantity Data**

At a minimum, the watershed or delineated wellhead area must be the same one(s) covered in the source water assessment. The utility shall maintain appropriate water quality data from the point of withdrawal from the source(s). In addition, water quality data from a variety of locations throughout the watershed or delineated wellhead area, where practicable, should be obtained to identify those areas that require additional scrutiny. Once these data are collected and analyzed, a subset of data and subwatersheds or wellhead zones should be described that effectively reflect real and potential problem areas. Sanitary surveys may be appropriate sources of further information pertinent to each watershed or delineated wellhead area under surveillance. The initial monitoring should be thorough and intense, leading ultimately to the development of a data-based framework for ongoing, time- and event-sensitive monitoring. Emerging issues, for example methyl tert-butyl ether, endocrine disruptors, perchlorate, hexavalent chromium, harmful algal blooms from nutrient enrichment, etc., should be carefully followed. As an adjunct part of a utility's standard operational activity, water quantity data should be gathered and analyzed in those areas where quantity issues are of real or potential concern. When deemed appropriate, the utility should work with partner organizations to establish water quality standards for its source water, which could encourage collaboration across multiple sectors. The sharing of water quality data among appropriate stakeholders and partners would further facilitate such collaborative processes to protect source water.

4.2.3   **Contaminant Sources and Land Use**

The utility shall maintain appropriate information and documentary support about known contaminant sources, land use activities, and other relevant information from the delineated area(s) of concern. Information about existing activities, controls, and management practices and their probable and real effectiveness in those areas shall be documented and maintained. The role of controls shall be adequately reviewed for appropriate use. Measurable results from these controls must be obtained. In addition, plans shall be in place to monitor future activities and development that may affect the watershed. Extreme events, such as floods, droughts and wildfire should be taken into considerations on source water characteristics. Utilities are encouraged to consider the regional long-term impact of climate change on source water quantity and quality. Water utilities should also consider working with other water users on the evaluation of alternative climate scenarios and mitigation and adaptation planning and investments. Some utilities may choose to conduct a risk or vulnerability analysis to address the various existing and anticipated threats to their source water.

4.2.4   Inventory of Regulations

The utility should compile a list of local, state or provincial, and federal regulations concerning land use management and other relevant activities that apply to the source water protection area. The utility should consider these regulations in the formulation of an action plan.

Rationale

Characterization and assessment of the source water and the land areas that affect it, along with a risk assessment analysis of the susceptibility of the source water to potential contamination, are critical to properly understanding the state of the environment and to developing appropriate goals and plans that will achieve the vision of the SWP program.

Characterization and Vulnerability Assessments

The characterization should, at a minimum, include the following:

- Delineation of the SWP area (surface water and/or wellhead)
- Baseline water quality and quantity information
- An inventory of the activities and potential contaminant sources in the SWP area (e.g., types of potential contaminant sources, the location of these sources, travel times from sources to public water supply intakes or wells)
- A risk assessment to determine the potential susceptibility of the water source to contamination from potential contaminant sources, as well as the risk inherent in that susceptibility (i.e., potential consequences)
- The risk assessment should be used to prioritize the threats from the various existing and potential contaminant sources

Under the 1996 Safe Drinking Water Act (SDWA) amendments, each state was required to develop source water assessment programs (SWAPs) and conduct source water assessments for all public water systems within their jurisdictions. Information collected and analyzed through these assessments was required to be made available to the public and was intended to be used voluntarily by local stakeholders to determine priorities for the establishment and implementation of successful SWP programs. Most of these source water assessments have been completed (USEPA 2006). However, because states had a limited time frame and limited resources, these baseline assessments were usually based on readily available data and therefore may be lacking local data of sufficient detail. In addition, given that many assessments were completed in the early 2000s, changes in land uses and other activities may have rendered the baseline assessments out of date. A USEPA manual is available to help guide stakeholders in updating their baseline assessments (USEPA 2006).

Additional data collection and analysis, along with other research activities, may be warranted to fill the information gaps and needs. A more comprehensive characterization process for source waters and SWP areas includes assessment of all relevant water quality data (including both regulated and unregulated contaminants), with associated consideration for potential spatial and temporal variability.

For example, in some cases rigorous approaches to delineating SWP areas could potentially make assessments scientifically defensible and credible for protection management. Starting with a copy of the existing SWAP report, a utility should evaluate the level of detail and sophistication of the methods used to delineate the SWP areas. A utility may then desire to perform more accurate delineations of SWP areas (e.g., using more sophisticated models and mapping), use higher spatial and temporal resolution for land-use data (e.g., retrieval of historical land-use data and modeling of build-out scenarios), and perform more detailed evaluations for susceptibility analyses (e.g., using contaminant fate and transport modeling and/or quantitative risk assessments).

The order of increasing accuracy and sophistication of the methods for delineating groundwater-based SWP areas is as follows (USEPA 1987, 2006):

- Arbitrary fixed radii
- Calculated fixed radii
- Simplified variable shapes
- Analytical methods (e.g., uniform flow equation)
- Hydrogeologic mapping
- Numerical flow/transport models (e.g., US Geological Survey's [USGS's] MODFLOW model)

For surface water systems, some states delineated SWP areas as entire watersheds upstream of the intakes up to the watershed/hydrologic boundaries or to the state boundaries. Given that some watersheds are very large, a segmentation approach can be used to identify smaller "critical areas" as high-priority areas for contaminant source inventories and susceptibility determinations. However, a utility may decide to expand existing critical areas because monitoring data and other information indicate that important potential contamination sources lie outside the currently defined critical area.

Generally one of the following four approaches is used to delineate critical areas for surface water intakes (USEPA 2006):

- An area defined by an arbitrary distance upstream or around the intake
- A stream time-of-travel distance upstream of an intake (stream only, not adjacent lands)
- A buffer-zone setback
- A stream time-of-travel area (stream plus watershed land)

More accurate and sophisticated methods are often associated with a higher degree of data requirements and costs. Consultation with local USGS district office staff, state geological survey office staff, university faculty members and research staff, and USEPA regional water staff can help in determining the level of accuracy and sophistication needed for the delineation process.

In certain cases, some of the desired data may not be available. Information gaps may need to be addressed before further activities can be completed.

Once a SWP area is appropriately delineated, information on land use and potential contaminant sources should be collected. Depending on the size of the source area being assessed and the level of detail desired, the characterization may involve the creation of maps and databases and the use of models and a geographic information system (GIS) to make collected data useable for the characterization process and for decision-making. Sources of information include the following:

- Zoning maps and build-out analyses
- Sanitary surveys
- Water quality monitoring data
- Citizen group activities
- National Pollutant Discharge Elimination System permits
- Resource Conservation and Recovery Act reporting
- USEPA's Drinking Water Mapping Application to Protect Source Waters system (www.epa.gov/sourcewaterprotection/dwmaps)

As part of the SWAP process, many states conducted only desktop exercises using a database of regulated activities to compile contaminant source inventories (CSIs), which sometimes lack the spatial resolution for assessing the risk associated with these contaminant sources. These original CSIs are snapshots in time and should be updated to reflect changing land use and activities. Here are a few ways to supplement, update, and/or enhance initial CSIs:

- Conduct a walk-through and/or a drive-through in the SWP area.
- Verify that the database information on regulated facilities and land uses is complete and accurate.
- Have sanitary survey technicians identify well/intake problems and additional potential contamination sources.
- Obtain information from various relevant reports and databases.
- Access USGS and state geological survey websites to determine the availability of water quality data and reports.
- Contact facilities identified through the CSI effort to verify their information.
- Conduct interviews and information searches to identify historic contamination events and locations.
- Add data with finer spatial resolution (e.g., from general land use to land parcel level).
- Replace and add data with more accurate locational attributes (e.g., through the use of a global positioning system).

Once supplemental and additional data are collected, a GIS can be used to manage, display, and analyze the data. Using GIS, a utility can gain a higher level of confidence of the identification of all significant potential contamination sources in the updated SWP area. Additional information that could make the characterization process more valuable in development of the SWP plan includes the following:

- Interviews, surveys, and tours of facilities, including reviews of facility regulatory permits and compliance records
- Interviews with municipal officials within the SWP area to further help identify current and potential contaminations sources
- Focused microbial source tracking investigations in areas where microbial contaminants have been previously detected to help identify the type and location of the contaminant source
- Identification of existing management practices and pollution prevention efforts

- Review of local and state land-use restrictions and regulations, planning documents, and permitting requirements associated with environmental and public health programs
- Identification and engagement of local citizens, organizations, and other stakeholders who are located within the SWP area or can affect source water quality

Once an understanding of the physical settings for the source water area is developed, it is important to assess the stakeholder, regulatory, and institutional attributes within the source water area where protection activities are desired. These characteristics require the identification of stakeholders' activities, roles, responsibilities, resources, and authority within the SWP area. An understanding of the relationships between the key players will ultimately lead to more effective planning and implementation. For example, permits, modifications of ordinances, or permission and participation from various parties to conduct activities on their lands may be needed to achieve various SWP goals. Furthermore, this may result in improved coordination of partnership activities, additional volunteer efforts, and potential funding opportunities. Therefore, engaging stakeholders and developing dialogues and long-term relationships with these entities over time is important and often productive in the long run.

The last step in the characterization process, which is an essential one, is to use the information gathered to perform a risk assessment analysis in order to determine the susceptibility of the source water to current or potential sources of contamination identified. This susceptibility analysis should prioritize the threats based on a variety of criteria, including likelihood of the current and potential contaminant sources to affect source water quality, the potential timeliness (e.g., continuous, frequent, infrequent, or rare) and magnitude of those potential water quality impacts, the potential adverse consequences caused by those impacts (e.g., to public health, regulatory compliance, treatment process integrity, aesthetic water quality), and other concerns.

There are many approaches to determining source water susceptibility. Typically, this involves combining data on potential contamination sources, settings and integrity of water intakes and wells, contaminant fate and transport, and ranking of contamination potentials. As an example, Louisiana conducted potential susceptibility analysis of its water supplies. A potential susceptibility analysis is a determination of the public water supply's susceptibility to contamination by significant potential sources identified within the SWP area. This consists of a sensitivity analysis, which includes factors inherent to the system and source water, and a vulnerability analysis, which includes the number and types of significant potential sources of contamination identified. Therefore, the potential susceptibility analysis combines a hydrogeologic or hydrologic sensitivity analysis with a vulnerability analysis within the delineated areas (Louisiana Department of Environmental Quality [LDEQ], no date a, no date b). The results of the analysis can be used as a basis for determining appropriate new protection measures or for re-evaluating current protection efforts.

Specifically, for groundwater systems, Louisiana's potential susceptibility analysis considers the following factors (LDEQ, no date a):

- Depth of the well (the shallower, the more sensitive)
- Age of the well (the older, the more sensitive)
- Average groundwater velocity in the aquifer in which the well is screened (the higher, the more sensitive)
- Soil recharge potential (the higher, the more sensitive)

Specifically, for surface water systems, Louisiana considers the following factors in determining sensitivity for these systems (LDEQ, no date b):

- Age of the intake (the older, the more sensitive)
- Length of streams in the SWP area (the longer, the more sensitive)
- Runoff factors—high precipitation, steep slope, low vegetative cover, and low soil permeability contribute to high runoff (the higher, the more sensitive)

Worksheets

Nine worksheets related to source water and SWP area characterization are included in Sec. 10. These include

- Worksheet B: Source Water and Source Water Protection Area Characterization
- Worksheet B-1: Delineating the Source Area of Concern
- Worksheet B-2: Water Quality Information
- Worksheet B-3: Contaminant Source Inventory Data
- Worksheet B-4: Land-Use Analysis
- Worksheet B-5: Physical Barrier Effectiveness Determination
- Worksheet B-6: Intake Structure
- Worksheet B-7: Filling Information Gaps and Needs
- Worksheet B-8: Analysis of Vulnerability/Susceptibility

Resources

AWWA (American Water Works Association). 2006. *Source Water Protection Best Management Practices* (2006 Participant Manual). Denver, Colo.: AWWA.

———. 2007. Manual of Water Supply Practices. *M50, Water Resources Planning.* 2nd ed. Chapter 11: Watershed Management and Groundwater Protection. Denver, Colo.: AWWA. http://www.awwa.org/store/productdetail.aspx?product id=6738.

CDM (Camp Dresser & McKee, Inc.). 2002. *Source Water Protection Reference Manual* (CD-ROM). Denver, Colo.: Water Research Foundation and AWWA.

Corona Environmental Consulting LLC. 2016 (in progress). *A Methodology for Locating and Managing Dynamic Potential Source Water Contaminant Data.* Denver, Colo.: Water Research Foundation. www.watersuite.com.

Davis, Chris (ed.). 2008. *Watershed Management for Drinking Water Protection.* Denver, Colo.: AWWA; St. Leonards, NSW, Australia: Australian Water Association.

Edge, Tom A. and Karl A. Schaefer (eds.). 2006. *Microbial Source Tracking in Aquatic Ecosystems: The State of Science and an Assessment of Needs.* NWRI Scientific Assessment Report Series No. 7. Burlington, Ontario: National Water Research Institute.

Long, S.C., and J.D. Plummer. 2004. Assessing Land Use Impacts on Water Quality Using Microbial Source Tracking. *Jour. AWRA*, 40(6):1433–1438.

Philadelphia Water Department. 2002a. *Belmont and Queen Lane Treatment Plants* (PWSID #1510001). Source Water Assessment Report. Philadelphia, Pa.: PWD. http://www.phillywatersheds.org/doc/Schuylkill_SWA_lowres.pdf.

———. 2002b. *Baxter Water Treatment Plant* (PWSID #1510001). Source Water Assessment Report. Philadelphia, Pa.: PWD. http://www.phillywatersheds.org/doc/Delaware_SWA.pdf.

Plummer, J.D., and S.C. Long. 2007. Monitoring Source Water for Microbial Contamination: Evaluation of Water Quality Measures. *Water Research*, 41(16):3716–3728.

Sturdevant Rees, Paula L., Sharon C. Long, Rebecca Baker, Daniel H. Bordeau, Routing Pei, and Paul K. Barten. 2006. *Development of Event-Based Pathogen Monitoring Strategies for Watersheds.* Denver, Colo.: Water Research Foundation.

USEPA (US Environmental Protection Agency). 1987. *Guideline for Delineation of Wellhead Protection Areas*. EPA 440-6-87/010. Washington, D.C.: USEPA, Office of Ground-Water Protection.

———. 1989. *Wellhead Protection Programs: Tools for Local Governments*. EPA 440-6-89-002. Washington, D.C.: USEPA, Office of Water.

———. 1991a. *Guide for Conducting Contaminant Source Inventories for Public Drinking Water Supplies: Technical Assistance Document*. EPA 570-9-91-014. Washington, D.C.: USEPA.

———. 1991b. *Managing Ground Water Contamination Sources in Wellhead Protection Areas: A Priority Setting Approach*. EPA 570-9-91-023. Washington, D.C.: USEPA, Office of Ground Water and Drinking Water. http://yosemite.epa.gov/water/owrcCatalog.nsf/9da204a4b4406ef885256ae 0007a79c7/62cba1aea27ee70b85256b0600723ed8!OpenDocument.

———. 1997a. *Delineation of Source Water Protection Areas: A Discussion for Managers*. Part 1: A Conjunctive Approach for Ground Water and Surface Water. EPA 816-R-97-012. Washington, D.C.: USEPA, Office of Water. http://yosemite.epa.gov /water/owrccatalog.nsf/e673c95b11602f2385256ae 1007279fe/7be186bf21 9d74d485256d83004fd80b?OpenDocument&Cart ID=1797-125417.

———. 1997b. *Guidelines for Wellhead and Springhead Protection Area Delineation in Carbonate Rocks*. EPA 904-B-97-003. Atlanta, Ga.: USEPA, Ground Water Protection Branch, Region 4.

———. 1997c. *State Source Water Assessment and Protection Programs: Final Guidance*. EPA 816-R-97-009. Washington, D.C.: USEPA Office of Water. https://semspub.epa.gov/work/HQ/124655.pdf.

———. 1999. *Protecting Sources of Drinking Water: Selected Case Studies in Watershed Management*. EPA 816-R-98-019, April 1999. Washington, D.C.: USEPA Office of Water.

———. 2001b. *Protecting and Restoring America's Watersheds: Status, Trends, and Initiatives in Watershed Management*. EPA 840-R-00-001. Washington, D.C.: USEPA Office of Water.

———. 2002c. USEPA Source Water Protection Practices Bulletins. http://cfpub. epa.gov/safewater/sourcewater/sourcewater.cfm?action=Publications&view =filter&document_type_id=103

———. 2003b. *Drinking Water Inspector's Field Reference: For Use When Conducting a Sanitary Survey of a Small Ground Water System*. EPA 816-R-03-023. Washington,

D.C.: USEPA, Office of Water. www.epa.gov/safewater/dwa/ sanitarysurvey /index.html.

———. 2003c. *Drinking Water Inspector's Field Reference: For Use When Conducting a Sanitary Survey of a Small Surface Water System.* EPA 816-R-03-022. Washington, D.C.: USEPA, Office of Water. www.epa.gov/safewater/dwa /sanitarysurvey/index.html.

———. 2005. *Microbial Source Tracking Guide Document.* EPA/600-R-05-064. Cincinnati, Ohio: USEPA Office of Research and Development. http://twri. tamu.edu/docs/bacteria-tmdl/epa%20mstguide%206-05.pdf.

———. 2006. *How-to Manual: Update and Enhance Your Local Source Water Protection Assessments.* Washington, D.C.: USEPA Office of Ground Water and Drinking Water. https://www.epa.gov/sites/production/files/2015-06 /documents/816k06004.pdf.

———. 2008. *Handbook for Developing Watershed Plans to Restore and Protect Our Waters.* EPA 841-B-08-002. Washington, D.C.: USEPA Office of Water. www.epa.gov/polluted-runoff-nonpoint-source-pollution/handbook -developing-watershed-plans-restore-and-protect.

———. 2008. *Handbook for Developing Watershed Plans to Restore and Protect Our Waters.* EPA 841-B-08-002. Washington, D.C.: USEPA Office of Water. www.epa.gov/polluted-runoff-nonpoint-source-pollution/handbook -developing-watershed-plans-restore-and-protect.

———. nd. *USEPA Source Water Web Guide.* http://permanent.access.gpo.gov /lps21800/www.epa.gov/safewater/protect/sources.html.

———. nd. *USEPA Watershed Central.* www.epa.gov/watershedcentral.

———. nd. *Drinking Water Mapping Application to Protect Source Waters (DWMAPS).* www.epa.gov/sourcewaterprotection/dwmaps.

Sec. 4.3   Source Water Protection Goals

Goals shall address water quality issues, such as public health and aesthetic concerns (e.g., taste and odor), and also may include other considerations, such as environmental stewardship, ecological balance, socioeconomic and political equity, trade-offs with competing policy objectives (such as transportation, housing, economic development, etc.), and others. Utilities are encouraged to develop their goals in cooperation with various stakeholders.

4.3.1   **Program Goals**

The utility should have written goals for the source water protection program. The utility shall include stakeholders in the development of the goals and shall document that involvement. At a minimum, the goals shall

- Address the specific problems or issues identified in the source water characterization element;

- Be expressed in terms that can be measured or otherwise evaluated in the future; and

- Meet or surpass existing and pending regulations, and provide the flexibility to incorporate future regulatory compliance.

The established goals could become the foundation of a strategy and/or strategic plan that would lead to the development of an Action Plan.

Rationale

The development of strategic goals that connect back to both the vision/mission statement and the source water area characterization and susceptibility analysis is an essential step in development of a SWP program. Such goals, often referred to as "key objectives," "critical business factors," or "critical success factors" in quality programs, become the basis for developing interim projects or activities for the program.

Some goals may be developed in response to specific problem areas identified during the characterization and susceptibility process, while other goals may be less technical or specific, such as "improve public education." Goals may address both current and potential future issues. Both internal and external stakeholders should be involved in development of the goals.

The goals should be prioritized to reflect the concerns of greatest importance and the areas and projects where success is most likely. To the extent practical, the goals should be SMART—specific, measurable, attainable, relevant, and timely). Goals should be realistic and obtainable in that there are no physical, environmental, institutional, financial, political, or social obstacles that are insurmountable.

Typically, the SWP program strategic goals are relatively high level and broad based and do not need to be modified for several years after being established. During implementation of SWP program projects and activities, additional information will be developed that may require the addition of new goals, modification of existing goals, or removal of existing goals once they have been achieved or deemed unnecessary or unachievable. Periodic review of the SWP program should include an evaluation of each goal as a measure of overall program success.

Worksheets

Worksheet C, Source Water Protection Goals, in Sec. 10 can be used when establishing goals.

Resources

Central Arkansas Water. 2007. *Lake Maumelle Watershed Management Plan*. Prepared by Tetra Tech, Inc. http://www.carkw.com/wp-content/uploads/2011/09/Lake_Maumelle_Watershed_Mgmt_Plan_May_07_reduced.pdf

National Research Council, Committee to Review the New York City Watershed Management Strategy. 2000. *Watershed Management for Potable Water Supply: Assessing the New York City Strategy*. Washington, D.C.: National Academy Press. www.nap.edu/catalog.php?record_id=9677#description.

PWD (Philadelphia Water Department). 2006. *The Schuylkill River Watershed Source Water Protection Plan (Belmont and Queen Lane Surface Water Intakes)*. Philadelphia, Pa.: PWD. http://www.phillywatersheds.org/doc/Schuylkill_SWPP_2006.pdf.

———. 2007. *The Delaware River Watershed Source Water Protection Plan (Baxter Water Treatment Plant Surface Water Intake)*. Philadelphia, Pa.: PWD. http://www.phillywatersheds.org/doc/Delaware_SWPP_2007.pdf.

USEPA (US Environmental Protection Agency). 2008. *Handbook for Developing Watershed Plans to Restore and Protect Our Waters*. EPA 841-B-08-002. Washington, D.C.: USEPA Office of Water. www.epa.gov/polluted-runoff-nonpoint-source-pollution/handbook-developing-watershed-plans-restore-and-protect.

Sec. 4.4   Action Plan

The action plan identifies required actions (management practices, statutory or regulatory changes, agreements, and so on) needed to mitigate existing and future threats to source water quality. It establishes priorities and sets a timetable to implement source water protection goals. The action items in the plan shall include the following:

4.4.1   Specific projects, programs, or other activities needed to achieve each of the source water protection goals shall be identified.

4.4.2   Specific projects, programs, and activities shall be prioritized, as appropriate, based on their likely effectiveness, availability of necessary resources, timing, stakeholder buy-in, political feasibility, and other considerations.

4.4.3   Necessary resources shall be identified, such as staff, funding, special expertise (e.g., police, health department and fire department), and provisions for obtaining them shall be included in the plan.

4.4.4   Potential barriers or obstacles to the action plan's implementation shall be acknowledged, and provisions for resolving them shall be incorporated into the plan.

4.4.5   Controls to monitor project/program progress, to document progress and successes, and to monitor funding or budgetary changes shall be identified.

4.4.6   Compliance With Regulatory Requirements - The utility should determine and document relevant local, state or provincial, federal, or other source water protection regulations that apply to their utility and its source water protection area. The utility shall comply with all applicable regulations for source water protection.

4.4.7   Security Planning and Implementation - The utility should have documentation that addresses security issues and describes, in detail, the response of personnel in the event of a security incident. Elements that address the protection of personnel and the water supply shall be included. The applicable vulnerability assessment shall be reviewed and consideration must be given to access control and other relevant security issues.

4.4.8   Emergency Preparedness and Response - The utility shall have documentation that describes emergency plans and provides specific directions to personnel in the event of an emergency. The program shall satisfy applicable regulatory requirements.

4.4.9   Health and Safety Management - The utility should document health and safety procedures that are designed to safeguard the employees and visitors engaged in operational activities pertaining to watershed management. The documentation may be specific to the source water(s) or part of a company-wide program.

Rationale

An action plan lays out a road map of activities to be conducted in order to achieve the desired SWP goals. The plan identifies selected actions (e.g., regulations, agreements, practices) to mitigate existing and future threats to source water quality. Protection of both current water supplies as well as potential new surface water sources or wells should be addressed. The action plan can also complement the broad sweep of community objectives, including economic development, quality of life, and other local issues.

The plan should prioritize actions as they relate to the goals developed for SWP based on characterization of the source and the susceptibility analysis. The prioritization analysis can include consideration, as appropriate, of the basis of the perceived risk from different contaminant sources, the likely effectiveness of different protection activities, availability of necessary resources, timing, stakeholder buy-in, political feasibility, the obstacles to success that exist for different contaminant sources and protection activities, and other considerations.

The action items may include such diverse SWP practices as land-use restriction and control regulations, public education, source water and early warning monitoring, structural best management practices for pollution control, and emergency response planning. Given that SWP is not mandated in most locales and that utilities typically have little or no regulatory authority or other legal mechanism to enforce specific practices, successful programs often must rely on their own merits and the voluntary actions of various stakeholders.

The action plan should include specifics as to how each action item will be accomplished and a timetable for implementation of each part of the plan. Necessary resources for accomplishing the plan, including staffing, special expertise, and the funding necessary to implement specific components of the plan (including budgets, identification of sources of funds, and a plan for securing that funding), should be identified. Potential problems and obstacles should be identified as fully as possible. Means and metrics for monitoring program effectiveness should also be defined. It is especially useful to designate someone to be responsible for championing the SWP program.

One challenging aspect of successful action plans is anticipating future needs. Future needs assessments should take into account potential future land-use patterns and regulatory activities. Use of a build-out analysis as part of the characterization process may help with assessment of the maximum potential change in land use.

As noted earlier, development of the SWP plan should be coordinated with regulatory agencies at all appropriate governmental levels. The utility staff should be aware not only of the regulations applicable to their utility but also of key regulations and ordinances that may affect other stakeholders and can potentially help the utility meet SWP goals. These could include, but are not limited to, zoning laws, land-use ordinances, state requirements for SWP activities, and activities associated with applicable total maximum daily loads.

An ongoing stakeholder education and involvement program should be included. Stakeholder involvement is typically essential for obtaining buy-in for the program and for leveraging resources and funding opportunities.

Security issues and emergency and contingency planning related to source waters should be addressed in order to the water system's overall resiliency (e.g., ANSI/AWWA J100 Risk and Resilience Management of Water and Wastewater Systems (2010). Security issues should be addressed in the assessment and planning process, and the utility should have written plans that describe the expected response of personnel in the event of a security incident. Elements that address the protection of personnel and the water supply need to be included. The source water vulnerability assessment can be used to help identify key security threats.

The program should include an emergency plan that details infrastructure and equipment in order to address accidents within the delineated area. Basically, all source waters are subject to contamination from a variety of potential sources, including accidental spills or releases of contaminants, impacts from major weather storms or other natural disasters (such as earthquakes and fire), and from deliberate terrorist activities. The utility should have documentation that describes plans and provides specific directions to personnel in the event of an emergency.

The program should also include contingency plans for locating and providing alternate drinking water supplies in the event of contamination. A good contingency plan ensures adequate planning for disasters, encourages reliability and consistency, contains measures to ensure continuity of operations, and creates uniform response protocols.

The health and safety of utility employees, contractors, and the general public are of utmost importance. Safety procedures, including appropriate safety gear, should be used to ensure the safety of all utility personnel and others who are working as part of the SWP program.

Worksheets

Three worksheets are provided in Sec. 10 for assistance in developing an action plan. These include

- Worksheet D: Source Water Protection Action Plans
- Worksheet D-1: Prioritization and Planning
- Worksheet D-2: Contingency Planning

Resources

AWWA (American Water Works Association). 2006. *Source Water Protection Best Management Practices* (2006 Participant Manual). Denver, Colo.: AWWA.

CDM (Camp Dresser & McKee, Inc.). 2002. *Source Water Protection Reference Manual* (CD-ROM). Denver, Colo.: Water Research Foundation and AWWA.

Center for Watershed Protection. 2000. Stormwater Manager's Resource Center (website). Ellicott City, Md.: Center for Watershed Protection. www.storm watercenter.net.

———. 2007. National Pollutant Removal Performance Database—Version 3. Ellicott City, Md.: Center for Watershed Protection. www.stormwaterok.net/ CWP%20Documents/CWP-07%20Natl%20Pollutant%20Removal%20 Perform%20Database.pdf

Central Arkansas Water. 2007. *Lake Maumelle Watershed Management Plan.* Prepared by Tetra Tech, Inc. http://www.carkw.com/wp-content/uploads/2011/09 /Lake_Maumelle_Watershed_Mgmt_Plan_May_07_reduced.pdf.

Davis, Chris (ed.). 2008. *Watershed Management for Drinking Water Protection.* Denver, Colo.: AWWA; St. Leonards, NSW, Australia: Australian Water Association.

Ernst, Caryn. 2004. *Protecting the Source: Land Conservation and the Future of America's Drinking Water.* Washington, D.C.: Trust for Public Land.

Ernst, Caryn, and Kelley Hart. 2005. *Path to Protection: Ten Strategies for Successful Source Water Protection.* Washington, D.C.: Trust for Public Land. http:// cloud.tpl.org/pubs/water_pathtoprotection.pdf.

Grayman, Walter M., Rolf A. Deininger, and Richard M. Males. 2001. *Design of Early Warning and Predictive Source-Water Monitoring Systems.* Denver, Colo.: Water Research Foundation.

Gullick, Richard W., Richard A. Brown, and David A. Cornwell. 2006. *Source Water Protection for Concentrated Animal Feeding Operations: A Guide for Drinking Water Utilities*. Denver, Colo.: Water Research Foundation.

Hopper, Kim, and Caryn Ernst. 2005. *Source Protection Handbook: Using Land Conservation to Protect Drinking Water Supplies*. Washington, D.C.: Trust for Public Land and Denver, Colo.: AWWA. www.tpl.org /source-protection-handbook.

Lampe, Les, Howard Andrews, Michael Barrett, Bridget Woods-Ballard, Peter Martin, Charles Glass, Neil Weinstein, and Chris Jeffries. 2004. *Post-Project Monitoring of BMPs/SUDs to Determine Performance and Whole Life Costs* (Report for Phase 1). Alexandria, Va.: Water Environment Research Foundation.

Oxenford, Jeff, et al. Source Water Protection Cost/Benefit Tool (http://www .swptool.org/index.cfm). Denver, Colo.: Water Research Foundation.

PWD (Philadelphia Water Department). 2006. *The Schuylkill River Watershed Source Water Protection Plan (Belmont and Queen Lane Surface Water Intakes)*. Philadelphia, Pa.: PWD. http://www.phillywatersheds.org/doc/Schuylkill _SWPP_2006.pdf.

———. 2007. *The Delaware River Watershed Source Water Protection Plan (Baxter Water Treatment Plant Surface Water Intake)*. Philadelphia, Pa.: PWD. http:// www.phillywatersheds.org/doc/Delaware_SWPP_2007.pdf.

Robbins, R.W., J.L. Glicker, D.M. Bloem, and B.M. Niss. 1991. *Effective Watershed Management for Surface Water Supplies*. Denver, Colo.: Water Research Foundation.

Rosen, Jeffrey S., Jose A.H. Sobrinho, Paul L. Freedman, and Wendy M. Larson. 2005. *Total Maximum Daily Loads (TMDLs) and Drinking Water Utilities*. Denver, Colo.: Water Research Foundation.

Schueler, Thomas R., and Heather K. Holland, Eds. 2000. *The Practice of Watershed Protection*. Ellicott City, Md.: Center for Watershed Protection.

Strange, Elizabeth M., Diana R. Lane, and Charles N. Herrick. 2009. *Utility Guidance for Mitigating Catastrophic Vegetation Change in Watersheds*. Denver, Colo.: Water Research Foundation.

USEPA (US Environmental Protection Agency). 1993. *Guidance Specifying Management Measures for Sources of Nonpoint Pollution in Coastal Waters*. EPA 840-B-92-002. Washington, D.C.: USEPA, Office of Water. www.

epa.gov/polluted-runoff-nonpoint-source-pollution/guidance-specifying -management-measures-sources-nonpoint.

———. 1999. *Protecting Sources of Drinking Water: Selected Case Studies in Watershed Management.* EPA 816-R-98-019, April 1999. Washington, D.C.: USEPA Office of Water.

———. 2003. *National Management Measures for the Control of Nonpoint Pollution from Agriculture.* EPA 841-B-03-004. Washington, D.C.: USEPA Office of Water. www.epa.gov/polluted-runoff-nonpoint-source-pollution /national-management-measures-control-nonpoint-source-0.

———. 2008. *Handbook for Developing Watershed Plans to Restore and Protect Our Waters.* EPA 841-B-08-002. Washington, D.C.: USEPA Office of Water. www. epa.gov/polluted-runoff-nonpoint-source-pollution/handbook-developing -watershed-plans-restore-and-protect.

———. 2013. *Federal Funding Opportunities for Source Water Protection.* www.epa .gov/sites/production/files/2015-04/documents/epa816k13001.pdf.

———. 2016. *Performance of Green Infrastructure, Databases and Summary Reports* (website) Washington, D.C.: USEPA. https://www.epa.gov/green-infrastructure/performance-green-infrastructure

———. nd. *USEPA Source Water Web Guide.* http://permanent.access.gpo.gov /lps21800/www.epa.gov/safewater/protect/sources.html.

———. nd. *USEPA Watershed Central.* www.epa.gov/watershedcentral.

———. nd. *USEPA National Agriculture Center.* www.epa.gov/agriculture.

———. nd. *Source Water Protection Funding.* www.epa.gov/sourcewaterprotection /funding-source-water-protection.

———. nd. *Catalog of Federal Funding Sources for Watershed Protection.* https:// ofmpub.epa.gov/apex/watershedfunding/f?p=fedfund:1.

Wright Water Engineers, Inc., and Geosyntec Consultants, Inc. 2010. International Stormwater BMP Database. www.bmpdatabase.org.

Sec. 4.5 Program Implementation

Implementation of the action plan is the key to a successful source water protection program. Responses to unexpected challenges and barriers to implementation of the action plan items should also be assessed in determining compliance with this standard. It is expected that some projects may be led or conducted by stakeholders other than the water utility or utilities.

As noted earlier, given that source water protection is a continuous, ongoing process that can span many years, it is not expected that a water utility would complete all aspects of its source water protection program within five or even 10 years from the initiation of the program. However, a water utility shall complete or demonstrate substantial progress in all six (6) elements of its source water protection program (as described in this standard), especially with regard to the implementation of key action items of its program. Performing source water assessment and developing a source water protection plan by themselves are not enough to satisfy the criteria of this standard. Completion or substantial implementation of action items is essential for a source water program to be deemed "in action" and for the generation of true results. Taking demonstrative steps in using this standard to achieve comprehensive source water protection is strongly encouraged.

4.5.1   The utility should, where appropriate, develop, promote, or implement a combination of voluntary and regulatory programs and sound practices such as

- Watershed planning
- Wellhead protection planning
- Land conservation
- Land use controls
- Contaminant source management
- Contingency planning
- Education and training
- Outreach and awareness programs
- Riparian buffers
- Green infrastructure and low-impact design standards
- Erosion and sediment control programs for construction projects
- Stormwater best management practices (BMPs)
- Agricultural best management incentives
- Watershed stewardship programs
- Responses to impacts from climate change and extreme events such as droughts and floods

- Security plan
- Health and safety plan
- Risk mitigation plan
- System vulnerability plan
- Operations plans

Rationale

As noted above, planning without implementation does not provide results, and without the implementation step, no actual protection is provided. Nonetheless, development of a comprehensive and implementable plan and use of an adaptive management approach to respond to unexpected challenges and barriers are integral to successful implementation of a SWP program. Accordingly, implementation success depends heavily on the groundwork performed up to this point. If a high-quality, comprehensive action plan is developed, the utility should have a good idea whether implementation of the plan will be successful.

When using a SWP program, a utility must be able to adequately gauge progress of the current implementation efforts and note how it has adapted to instances of adversity and challenge. The ability to work within and overcome institutional obstacles is a hallmark of a SWP program's viability. Forging new partnerships as well as strengthening traditional alliances are also critical.

Worksheets

Worksheet E, Program Implementation, and the six supplementary worksheets in Sec. 10 walk the user through important considerations for implementing a SWP program. Supplementary worksheets include:

- Worksheet E-1: Assessing Key Milestones
- Worksheet E-2: Roles and Responsibilities
- Worksheet E-3: Resources
- Worksheet E-4: Water Quality Monitoring
- Worksheet E-5: Biological/Habitat Monitoring
- Worksheet E-6: Stakeholder and Public Relations

Resources

USEPA (US Environmental Protection Agency). 1999. *Protecting Sources of Drinking Water: Selected Case Studies in Watershed Management.* EPA 816-R-98-019, April 1999. Washington, D.C.: USEPA Office of Water.

————. 2000. *Watershed Success Stories—Applying the Principles and Spirit of the Clean Water Action Plan.* Washington, D.C.: USEPA.

————. 2008. *Handbook for Developing Watershed Plans to Restore and Protect Our Waters.* EPA 841-B-08-002. Washington, D.C.: USEPA Office of Water. www.epa.gov/polluted-runoff-nonpoint-source-pollution/handbook-developing-watershed-plans-restore-and-protect.

Sec. 4.6   Evaluation and Revision

Source water protection programs shall be periodically evaluated and revised in response to changes in the area of source water delineation, new data or information, new regulatory initiatives, changes in local priorities, actual performance of implemented programs, and so forth.

4.6.1   Evaluation Procedures

The source water protection program should include provisions for periodically reviewing and, if necessary, modifying the utility's source water protection vision, characterization, goals, and implementation elements. This adaptive management approach (as a step in the process) is intended to measure the accomplishment or completion of projects, programs, and activities identified in the action plan. It also aims to identify gaps and shortcomings in the program for making future improvements.

Rationale

Any type of administrative program requires periodic (or continuous) evaluation and revision. A good SWP program includes provisions for reviewing, benchmarking, and, if necessary, modifying the SWP vision, characterization, goals, action plan, and implementation elements. These tasks should be conducted periodically and also in response to changes in the source water area, changes in contaminant sources, new data, performance of implemented programs, and other changes. This step of the process is intended to measure the accomplishment or completion of projects, programs, and activities identified in the action plan and also to recognize obstacles to success and identify ways to overcome those obstacles.

Methods for measuring the effectiveness of as many facets of the SWP program as possible must be established. In particular, measurement of the effectiveness of the specific action plan items implemented is important (e.g., efficacy of specific BMPs). In addition, the program should have a mechanism to evaluate and respond to new data collected, to changes in the watershed over time, and to the measured performance of implemented action items. The evaluation and revision facets of the SWP program should be designed with the idea that SWP is an iterative and interactive process, enabling the SWP plan to be a living document that continuously undergoes improvement.

Worksheet

Worksheet F, Program Evaluation and Revision, in Sec. 10 can be used for assistance with SWP program evaluation and revision.

Resources

USEPA (US Environmental Protection Agency). 2006. *How-to Manual: Update and Enhance Your Local Source Water Protection Assessments.* Washington, D.C.: USEPA Office of Ground Water and Drinking Water. https://www.epa.gov/sites/production/files/2015-06/documents/816k06004.pdf.

———. 2008. *Handbook for Developing Watershed Plans to Restore and Protect Our Waters.* EPA 841-B-08-002. Washington, D.C.: USEPA Office of Water. www.epa.gov/polluted-runoff-nonpoint-source-pollution/handbook-developing-watershed-plans-restore-and-protect.

SECTION 5: VERIFICATION

It is essential that the utility maintain adequate records and documents of its source water protection program to verify compliance with this standard. Such records shall include, but not be limited to, executed resolutions and recorded minutes of the utility's governing body; summaries or minutes of relevant public hearings, advisory committee, or stakeholder meetings; technical studies, monitoring data, memoranda, or other documents that support the delineation, goal-setting, characterization, and implementation elements of the source water protection program.

Rationale

Effective document management is essential to the successful and efficient progress of any utility program. Standard operating procedures that cover creation, approval, publication, distribution, revision, archiving, and destruction of electronic and paper documents and records should be developed. Also, someone should be formally designated as responsible for the management of all documents and records.

Documentation (e.g., source water protection plans) should be approved, reviewed, and revised as necessary on a regular schedule and be available at all relevant points of use within the organization. Procedures need to be in place so records are maintained as long as necessary and only current documents are accessible. Unfortunately, it is not unusual to start working with a document only to find that it has been superseded by a more recent update. Once a document has been replaced, a procedure for recalling and replacing the previous edition must be in place and implemented.

Proper documentation of utility and employee goals helps to provide clear direction as to what tasks need to be accomplished and in what time frame. If goals are not documented, it is easy to lose sight of the big picture, miss some key priority actions, and lose valuable time attending to unimportant tasks.

Worksheet

Worksheet G, Verification and Record Keeping, provides guidance on what documentation is required to help verify that the requirements of the standard have been met.

SECTION 6: GLOSSARY OF ABBREVIATIONS

ACWA	Association of Clean Water Administrators
ANSI	American National Standards Institute
ASDWA	Association of State Drinking Water Administrators
ASIWPCA	Association of State and Interstate Water Pollution Control Administrators
AWWA	American Water Works Association
BMP	best management practice
CAW	Central Arkansas Water

CCL	Contaminant Candidate List
CCME	Canadian Council of Ministers of the Environment
CSI	contaminant source inventory
CWA	Clean Water Act
CWP	Center for Watershed Protection
DRBC	Delaware River Basin Commission
EWS	early warning system
FIFRA	Federal Insecticide, Fungicide, and Rodenticide Act
GIS	geographic information system
GWPC	Ground Water Protection Council
GWUDI	groundwater under the direct influence of surface water
LDEQ	Louisiana Department of Environmental Quality
LT2ESWTR	Long-Term 2 Enhanced Surface Water Treatment Rule
MCL	maximum contaminant level
MST	microbial source tracking
NEIWPCC	New England Interstate Water Pollution Control Commission
NGO	nongovernmental organization
NPDES	National Pollutant Discharge Elimination System
NRC	National Research Council
NRCS	Natural Resources Conservation Service
PAC	Policy Advisory Council
PADEP	Pennsylvania Department of Environmental Protection
PBE	physical barrier effectiveness
PSC	potential source of contamination
PWD	Philadelphia Water Department
PWSID	public water system identification number
RCRA	Resource Conservation and Recovery Act
SAN	Schuylkill Action Network
SDWA	Safe Drinking Water Act
SOP	standard operating procedure
SPCC	spill prevention, control, and countermeasures
SUDS	sustainable urban drainage system
SWAP	source water assessment program
SWP	source water protection
TMDL	total maximum daily load

TOC	total organic carbon
USDA	US Department of Agriculture
USEPA	US Environmental Protection Agency
USGS	US Geological Survey
WATERS	Watershed Assessment, Tracking & Environmental Results
WRF	Water Research Foundation
WTP	water treatment plant

SECTION 7: RESOURCES

Supporting Organizations and Stakeholders

Effective source water protection (SWP) programs typically involve collaboration among a variety of stakeholders, including utilities; municipal, county, state, and federal governments and agencies; land-use planners; industries; agricultural interests; landowners and developers; emergency responders; watershed organizations; environmental and citizen groups; the media; and other organizations. Examples of select key organizations that have played an important role in fostering SWP activities on a national or large regional scale are listed below. Websites for many of these organizations are provided, along with web links for numerous helpful SWP resources.

US Federal Government Agencies

The US Environmental Protection Agency (USEPA) Office of Water includes the Office of Ground Water and Drinking Water; Office of Wetlands, Oceans, and Watersheds; Office of Wastewater Management; and Office of Science and Technology.

The US Department of the Interior includes the US Geological Survey (USGS), which produces water quality monitoring data and conducts research related to groundwater and surface water, and the Bureau of Land Management, National Park Service, and US Fish and Wildlife Service.

The US Department of Agriculture (USDA) has several programs that provide helpful information and services, including the Natural Resources Conservation Service, which has developed databases and computer programs related to the costs of agricultural best management practices (BMPs); the National Institute of Food and Agriculture (formerly the Cooperative State Research, Education, and

Extension Service); Agricultural Research Service; Farm Service Agency; Conservation Effects Assessment Project; and the US Forest Service.

State Environmental and Public Health Agencies

Some of these agencies are represented by various organizations such as the Association of State Drinking Water Administrators (ASDWA), Association of Clean Water Administrators (formerly the Association of State and Interstate Water Pollution Control Administrators), and Ground Water Protection Council (GWPC).

The Source Water Collaborative

Currently comprised of 26 federal, state, and local organizations and led by the USEPA, the Source Water Collaborative (http://sourcewatercollaborative.org) was formed in 2006 to further the goal of protecting drinking water sources. Recognizing that available resources are extremely limited, authorities are split, and the stakeholders who can actually protect source waters are diffuse, the Source Water Collaborative's intention was to collaborate among organizations to combine their various strengths and tools. While each organization individually promotes implementation of SWP in their overall mission, they also recognize the synergy of coordinated actions and the need to leverage each other's resources in order to increase the chances for success. Key Source Water Collaborative members include the USEPA, USDA, AWWA, ASDWA, American Planning Association, GWPC, National Rural Water Association, National Association of Counties, Groundwater Foundation, and Trust for Public Land. Web links to a variety of SWP tools are available on their websites.

River Basin Commissions

Several river basin commissions have been formed in the United States via interstate compacts to plan and implement water resources strategies on a large watershed scale across state boundaries. Examples include the Ohio River Valley Water Sanitation Commission, Susquehanna River Basin Commission, Delaware River Basin Commission, and Interstate Commission on the Potomac River Basin.

Watershed Associations

Various regional watershed coalitions and information centers are located throughout the United States. Some provide linkages to watershed associations within their respective states or regions. Examples include the Missouri Watershed Information Network and the Massachusetts Watershed Coalition.

Environmental Finance Centers

USEPA funds eight university-based environmental finance centers in the United States that provide education and other services related to a variety of environmental finance issues including SWP activities. They deliver targeted technical assistance to and partner with states, tribes, local governments, and the private sector in providing innovative solutions to help manage the costs of environmental financing and program management. They have also provided assistance for several SWP case studies, including analysis of funding avenues and resulting costs and benefits.

Land Trust Organizations

There are numerous organizations devoted to land conservation and preservation in the United States, including the Trust for Public Land, the Nature Conservancy, and the Land Trust Alliance (a national convener, strategist, and representative of more than 1,600 US land trusts).

Other Nongovernmental and Non-US Organizations

Numerous key nongovernmental organizations exist that promote SWP. These include AWWA (including AWWA's volunteer Source Water Protection Committee that developed Standard G300 and this guidebook), Water Research Foundation, Water Environment Federation, Water Environment & Reuse Foundation, American Water Resources Association, the Conservation Technology Information Center, National Environmental Services Center, Rural Community Assistance Partnership, and the Center for Watershed Protection.

Canadian organizations include the environmental agencies of the various provinces (e.g., Ontario Ministry of the Environment and Climate Change). On a provincial level, Conservation Ontario is an umbrella organization that represents a network of 36 conservation authorities throughout Ontario, Canada. These conservation authorities, created in 1946 by an act of the provincial legislature, are mandated to ensure the conservation, restoration, and responsible management of Ontario's water, land, and natural habitats through programs that balance human, environmental, and economic needs.

Select Websites Related to Source Water Protection

US Environmental Protection Agency

- *USEPA Source Water Protection website:* www.epa.gov/sourcewaterpro-tection. This website contains links to a variety of helpful information, including basic information on source waters, local protection activities, partnerships for protection, tools and training, guidance manuals, case studies, and links to many other relevant websites.
- *USEPA Office of Ground Water and Drinking Water:* www.epa.gov/ogwdw
- *USEPA Office of Water:* www.epa.gov/ow
- *USEPA Office of Wetlands, Oceans, and Watersheds:* www.epa.gov/owow
- *USEPA Source Water Web Guide:* http://permanent.access.gpo.gov/lps21800/www.epa.gov/safewater/protect/sources.html. This annotated web guide is a selected collection of available SWP tools, primarily focused on source water resources either created by USEPA or which USEPA has supported through grants.
- *Source Water Protection Practices Bulletins:* http://cfpub.epa.gov/safewater/sourcewater/sourcewater.cfm?action=Publications&view=filter&document_type_id=103. This site includes a series of 13 bulletins that address SWP for 13 types of contamination sources.
- *Watershed Academy Web:* www.epa.gov/watertrain; www.epa.gov/watershedacademy/online-training-watershed-management#themes. Developed by the USEPA Office of Wetlands, Oceans, and Watersheds, this website offers a variety of self-paced training modules that represent a basic and broad introduction to the watershed management field. The various modules include an introduction and overview, and there are modules that cover watershed ecology, watershed change, analysis and planning, management practices, and community and social aspects, and water law.
- *Drinking Water Mapping Application to Protect Source Waters (DWMAPS):* www.epa.gov/sourcewaterprotection/dwmaps. This is an online mapping tool that can be used when conducting a source water assessment, in particular, through identification of potential sources of contamination in locations defined by users.
- *Surf Your Watershed:* http://cfpub.epa.gov/surf/locate/index.cfm. Developed by the USEPA Office of Wetlands, Oceans, and Watersheds, this website provides the general public and watershed practitioners geographically

relevant information about each watershed in the United States. Resources include a tool to locate any watershed anywhere in the United States, assessments of watershed health, links to related USGS data, state and tribal information, a watershed atlas, information on how to "adopt" your watershed, a listing of citizen groups active in the watershed, and links to websites of other environmental-related organizations active within each specific watershed.

- *WATERS* (Watershed Assessment, Tracking & Environmental ResultS): www.epa.gov/waters. This powerful website provides an integrated information system to enable the user to easily obtain information and data for the nation's surface waters, which was previously available only from several independent and unconnected databases. Among the several available tools is EnviroMapper for Water, which is a web-based geographic information system (GIS) application that dynamically displays information about bodies of water in the United States.

- *Clean Water Act (CWA):* https://www.epa.gov/laws-regulations/summary-clean-water-act including the National Pollutant Discharge Elimination System (NPDES) and total maximum daily loads (TMDL) programs.

- *Safe Drinking Water Act (SDWA):* www.epa.gov/safewater/sdwa/index.html

- *Resource Conservation and Recovery Act (RCRA):* www.epa.gov/rcra/resource-conservation-and-recovery-act-rcra-overview

- *Comprehensive Environmental Response, Compensation and Liability Act (CERCLA), or Superfund):* www.epa.gov/superfund/index.htm

- *Frank R. Lautenberg Chemical Safety for the 21st Century Act:* www.epa.gov/assessing-and-managing-chemicals-under-tsca/frank-r-lautenberg-chemical-safety-21st-century-act (this law updates the Toxic Substances Control Act)

- *Federal Insecticide, Fungicide, and Rodenticide Act (FIFRA):* www.epa.gov/laws-regulations/summary-federal-insecticide-fungicide-and-rodenticide-act

- *Agriculture (formerly the USEPA National Agriculture Compliance Assistance Center):* www.epa.gov/agriculture. This site includes information on animal production practices, BMPs, research, regulations, and other pertinent topics.

- Animal Feeding Operations (AFOs): https://www.epa.gov/npdes/animal-feeding-operations-afos

- *Comprehensive Nutrient Management Planning information:* www. nrcs.usda.gov/wps/portal/nrcs/detail/pa/technical/ecoscience/ nutrient/?cid=nrcs142p2_018164
- *Source Water Protection Funding:* www.epa.gov/sourcewaterprotection/ funding-source-water-protection. This USEPA Office of Ground Water and Drinking Water website contains a variety of information related to potential funding sources for SWP activities.
- *Catalog of Federal Funding Sources for Watershed Protection:* https:// ofmpub.epa.gov/apex/watershedfunding/f?p=fedfund:1. This USEPA Office of Water website includes a searchable database of federal financial assistance sources (e.g., grants, loans, and cost-sharing) available to fund a variety of watershed protection projects. The use can search by either subject matter criteria or based on words in the title of the funding program. Criteria searches include the funding agencies, type of eligible organization (e.g., nonprofit groups, private landowner, state, business), type of assistance sought (grants or loans), and keywords (e.g., agriculture, wildlife habitat).
- *Nonpoint Source News-Notes:* www.epa.gov/newsnotes. This site presents an extensive periodic news report on information and resources related to the condition of the water-related environment, the control of nonpoint sources of water pollution, and the ecological management and restoration of watersheds. One can subscribe to receive notification via e-mail, and all back issues are available online.

US Department of Agriculture

- *Farm Bill:* www.usda.gov/wps/portal/farmbill2008?navid= FARMBILL2008
- *Source Water Protection Program:* www.fsa.usda.gov/programs-and-services/ conservation-programs/source-water-protection/index. This is a joint project with the US Department of Agriculture Farm Service Agency and the National Rural Water Association. Collaborative teams are formed to create a rural SWP plan, and voluntary actions that farmers and ranchers can take to prevent source water pollution are identified.
- *Watershed and Flood Prevention Operations Program:* www.nrcs.usda.gov/ wps/portal/nrcs/main/national/programs/landscape/wfpo. This program provides technical and financial assistance to states, local governments, and tribes (project sponsors) for planning and implementing authorized

watershed project plans for the purpose of watershed protection, water quality improvements, soil erosion reduction, sediment control, water management, flood mitigation, and other activities.

- *Animal Feeding Operations (AFOs):* http://www.nrcs.usda.gov/wps/portal/ nrcs/main/national/plantsanimals/livestock/afo. This website from the USDA Natural Resources Conservation Service (NRCS) has web links for guidance on how to control pollution from AFOs. Included is guidance on comprehensive nutrient management plans.

- *Water Quality:* www.nrcs.usda.gov/wps/portal/nrcs/main/national/water/ quality/. This website with links to NRCS information about water management, water quality, and watersheds.

- *National Conservation Practice Standards–NHCP (NRCS):* www.nrcs.usda. gov/wps/portal/nrcs/main/national/technical/cp/ncps. This site include a list of National Conservation Practice Standards.

- *Agricultural Research Service National Research Programs and Information:* www.ars.usda.gov/research/programs.htm. See the programs within the section on Natural Resources and Sustainable Agricultural Systems (www. ars.usda.gov/natural-resources-and-sustainable-agricultural-systems).

- *Water and Agriculture Information Center:* www.nal.usda.gov/waic. This site provides electronic access to information on water and agriculture. The center collects, organizes, and communicates the scientific findings, educational methodologies, and public policy issues related to water and agriculture. This is part of the National Agricultural Library and includes links to numerous articles.

US Geological Survey

USGS Water Resources Programs (water quality monitoring and research programs): http://water.usgs.gov/programs.html

USGS Water-Quality Information Pages: http://water.usgs.gov/owq

USGS Groundwater Information Pages: http://water.usgs.gov/ogw

USGS Surface Water Information Pages: http://water.usgs.gov/osw

Animal Feeding Operations—Effects on Confined Animal Feeding Operations on Hydrologic Resources and the Environment: http://water.usgs.gov/owq/AFO

Animal Feeding Operations: http://toxics.usgs.gov/regional/emc/animal_ feeding.html. This site includes information on the occurrence of veterinary pharmaceuticals and antibiotics in animal feeding operation wastes.

- *Water-Quality Information—Microbial Source-Tracking and Detection Techniques:* http://water.usgs.gov/owq/microbial.html. Links to other sites and references such as USGS publications

Other Organizations

- *American Water Works Association (AWWA):* www.awwa.org
- *Association of State Drinking Water Administrators (ASDWA):* www.asdwa.org
- *Association of Clean Water Administrators (ACWA):* www.acwa-us.org
- *American Water Resources Association (AWRA):* www.awra.org
- *Association of Metropolitan Water Agencies (AMWA):* http://www.amwa.net
- *Center for Watershed Protection:* www.cwp.org
- *Groundwater Foundation:* www.groundwater.org
- *Ground Water Protection Council (GWPC):* www.gwpc.org
- *New England Interstate Water Pollution Control Commission (NEIWPCC):* www.neiwpcc.org
- *National Rural Water Association (NRWA):* www.nrwa.org
- *Source Water Collaborative:* http://sourcewatercollaborative.org
- *Trust for Public Land:* www.tpl.org
- *Water EUM (Effective Utility Management):* http://www.watereum.org
- *Water Research Foundation:* www.waterresearchfoundation.org

BMP Databases

- *International Stormwater BMP database:* www.bmpdatabase.org (Wright Water Engineers, Inc. and Geosyntec Consultants, Nd.)
- *Stormwater Manager's Resource Center:* www.stormwatercenter.net (Center for Watershed Protection 2000)
- *National Pollutant Removal Performance Database (Version 3):* http://www.stormwaterok.net/CWP%20Documents/CWP-07%20Natl%20Pollutant%20Removal%20Perform%20Database.pdf (Center for Watershed Protection 2007)
- *Performance of Green Infrastructure, Databases and Summary Reports:* www.epa.gov/green-infrastructure/performance-green-infrastructure (USEPA 2016)

SECTION 8: ANNOTATED BIBLIOGRAPHY FOR SELECT INFORMATION SOURCES

The following table provides brief descriptions of various reference sources that provide information and tools to support source water protection (SWP) activities. Many of these sources will lead the reader to numerous other helpful references and websites.

This review focuses primarily on publications from national-level organizations. A complete review of all such publications is not possible here, but some of the more useful, popular, and readily available references are listed. Many state agencies and regional and local organizations have also published numerous excellent guidance documents and other helpful information. The reader is encouraged to consult with their state and local environmental officials and perform Internet searches to find this type of information published in or written about their own area.

Reference Title (Citation)	Author and Description
American Water Works Association: www.awwa.org	
ANSI/AWWA Standard G300-14 for Source Water Protection (AWWA 2014)	AWWA provides the definitive standard for water utility SWP programs. ANSI/AWWA G300-14, Source Water Protection outlines the six primary components of successful SWP programs and provides the requirements for the standard plus guidance to help meet it. Available for purchase at: www.awwa.org/store/productdetail.aspx?productid=39840706.
Source Water Protection Best Management Practices—2006 Participant Manual (AWWA 2006)	This training manual supports a two-day course on SWP offered periodically by AWWA. It covers developing a SWP plan, performing source water assessments and determining source water susceptibility, developing emergency plans, using public participation and education as a vital component of SWP, and securing funding for SWP projects.
Watershed Management for Drinking Water Protection (Davis 2008)	This document describes practices for evaluating risks to water quality and options for reducing risks within a watershed, ways to secure funding for SWP activities, stakeholder involvement and partnerships, land-use planning, surveillance, monitoring, recreational access, on-site wastewater treatment, groundwater protection, communication, and incident management. Each chapter includes a section on best practices, and case studies are provided.
US Environmental Protection Agency: www.epa.gov	
USEPA websites for SWP information	Source Water Protection website: This site contains links to a variety of helpful information, including basic information on source waters, local protection activities, partnerships for protection, tools and training, guidance manuals, case studies, and links to many other relevant websites. www.epa.gov/sourcewaterprotection
	Source Water Web Guide: This annotated guide includes a selected collection of available SWP tools, primarily focusing on source water resources either produced by USEPA or supported by USEPA through grants. http://permanent.access.gpo.gov/lps21800/www.epa.gov/safewater/protect/sources.html
	Watershed Central. This is a web-based system for organizing USEPA's information and tools using an integrated watershed management framework. The goal is to help watershed organizations and others find key resources to protect their local watershed. Users can find environmental data, watershed models, local organizations, and guidance documents. The site also contains links to watershed technical resources and funding, mapping applications to help find information specific to named watersheds, and a Watershed Wiki that watershed practitioners can use to collaborate. www.epa.gov/watershedcentral

Reference Title (Citation)	Author and Description
US Environmental Protection Agency: www.epa.gov	
USEPA websites for SWP information (*continued*)	National Agriculture Center: This site includes information on animal production practices, BMPs, research, regulations, and other pertinent topics.
	www.epa.gov/agriculture
	Section 319, Nonpoint Source Success Stories: This site includes more than 100 case studies that discuss ways to improve water quality for surface waters that have been classified on the Clean Water Act 303(d) list of impaired waters.
	www.epa.gov/polluted-runoff-nonpoint-source-pollution/nonpoint-source-success-stories
USEPA websites for SWP funding	Source Water Protection Funding, a USEPA Office of Ground Water and Drinking Water website: This site contains information related to potential funding sources for SWP activities. www.epa.gov/sourcewaterprotection/funding-source-water-protection
	Catalog of Federal Funding Sources for Watershed Protection: This Office of Water website includes a searchable database of federal financial assistance sources (grants, loans, cost-sharing) available to fund a variety of watershed protection projects. The database can be searched by either subject matter criteria or based on words in the title of the funding program. Criteria searches include the funding agencies, type of eligible organization (e.g., nonprofit groups, private landowner, state, business), type of assistance sought (grants or loans), and keywords (e.g., agriculture, wildlife habitat). https://ofmpub.epa.gov/apex/watershedfunding/f?p=fedfund:1
Federal Funding Opportunities for SWP (USEPA 2013)	This six-page brochure lists federal funding sources for a variety of SWP and environmental restoration projects. www.epa.gov/sites/production/files/2015-04/documents/epa816k13001.pdf
Drinking Water Mapping Application to Protect Source Waters	This is an online mapping tool that helps users conduct source water assessments, in particular, through identification of potential sources of contamination in locations defined by the user. www.epa.gov/sourcewaterprotection/dwmaps
Annotated Bibliography of Source Water Protection Materials (USEPA 2003a)	Available in both CD (816-F-03-010) and print form (EPA 816-C-03-003; 73 pages), this extensive bibliography on SWP was gathered from a variety of sources and includes fact sheets, reports, guides, outreach materials, regulations, web pages, and videos. Organized by subject area, the CD includes a search function to locate information. The printed report is available at www.sourcewaterpa.org/wp-content/uploads/2010/10/Annotated-Bibliography-of-Source-Water-Protection-Materials.pdf
Watershed Academy	The USEPA Office of Wetlands, Oceans, and Watersheds provides a variety of self-paced training modules that represent a basic and broad introduction to the watershed management field (www.epa.gov/watertrain). The modules include an introduction/overview, with some covering watershed ecology, watershed change, analysis and planning, management practices, community and social aspects, and water law. www.epa.gov/watershedacademy/online-training-watershed-management#themes
Source Water Protection: Best Management Practices and Other Measures for Protecting Drinking Water Supplies (USEPA 2002b)	This extensive slide presentation covers the basics of SWP and includes detailed notes for many of the slides. https://cfpub.epa.gov/watertrain/pdf/swpbmp.pdf
Source Water Protection Practices bulletins (USEPA 2002c)	This series of 13 bulletins addresses SWP for 13 types of contamination sources. http://cfpub.epa.gov/safewater/sourcewater/sourcewater.cfm?action=Publications&view=filter&document_type_id=103
National Source Water Contamination Prevention Strategy—Seventh Draft for Discussion (USEPA 2001a)	This publication provides an overview of the challenges to preserving and protecting water sources; providing a vision for contamination prevention; presenting the mission and goals of SWP and describing a strategic approach for meeting SWP objectives; and establishing performance measures. The strategy is presented in four parts, including long-term goal, vision, and building blocks; shorter-term strategic approach; measuring progress; and emerging issues. This ambitious strategy was never completed or adopted by USEPA, and while many of the proposed USEPA actions have not yet been performed, some of the metric measurements identified in the strategy have since been used. It does provide numerous ideas and strategies for developing a national road map for source water protection.
	http://permanent.access.gpo.gov/lps21800/www.epa.gov/safewater/protect/strateg7.pdf
Handbook for Developing Watershed Plans to Restore and Protect Our Waters (USEPA 2008)	This handbook supplements the many other watershed planning guides. It is generally more detailed than other guides with regard to quantifying existing pollutant loads, developing estimates of the load reductions required to meet water quality standards, developing effective management measures, and tracking program progress. Its approach uses the six elements used in AWWA G300, Standard for Source Water Protection. www.epa.gov/polluted-runoff-nonpoint-source-pollution/handbook-developing-watershed-plans-restore-and-protect

Reference Title (Citation)	Author and Description
US Environmental Protection Agency: www.epa.gov	
State Source Water Assessment and Protection Programs Guidance: Final Guidance (USEPA 1997c)	This document provides instructions for completing source water assessments (SWAP reports). Chapters include the following: Overview of Source Water Assessment and Protection and the SDWA; Final Guidance for State SWAPs; Tools for State Source Water Protection Implementation Including Petition Programs and the Drinking Water State Revolving Fund; Relationship Between Source Water Assessments, Source Water Protection Programs, and the Public Water Supply Supervision Program; and Coordination of Source Water Assessments, Source Water Protection Programs, and Other EPA and Federal Programs.
How-to Manual: Update and Enhance Your Local Source Water Protection Assessments (USEPA 2006)	This manual provides guidance on how to update the various components of a source water assessment. Most state-generated SWAP reports have been completed. However, because states had a limited time frame and limited resources, most of these baseline assessments were based on readily available data and therefore may not include sufficient detail. In addition, given that many assessments were completed in the early 2000s, there may have been changes in land uses and other activities that would render the baseline assessment currently incomplete. This manual provides guidance on how to improve a baseline assessment using more detailed information and data and more accurate assessment methods. www.epa.gov/sourcewaterprotection/how-manual-update-and-enhance-your-local-source-water-protection-assessments
Protecting and Restoring America's Watersheds: Status, Trends, and Initiatives in Watershed Management (USEPA 2001b)	This report describes various successes and ongoing barriers to the effective use of watershed-based approaches. The report reflects the input and experience of nearly all federal agencies involved in the management of watersheds, as well as input from state agencies, tribes, watershed groups, academicians, nonprofit organizations, and private citizens. It explores the successes of selected case studies and evaluates programs and partnerships representative of the larger national efforts underway to move stakeholders toward a watershed management approach. The report also discusses areas that many stakeholders believe still need improvement.
Microbial Source Tracking Guide Document (USEPA 2005)	This guide provides a comprehensive, interpretive analysis of microbial source tracking (MST), especially as it is used for various water quality evaluations (e.g., drinking water source assessment and protection, public health issues, beach closures, microbial risk management, and ecosystem restoration), total maximum daily load–related activities, and cases where accurate identification of fecal pollution is required to implement reliable management practices. It includes descriptions of various MST approaches, data collection tools, data analysis procedures, method applications, and performance standards, as well as various assumptions and limitations associated with MST. http://twri.tamu.edu/docs/bacteria-tmdl/epa%20mstguide%206-05.pdf
National Management Measures to Control Nonpoint Source Pollution from Agriculture (USEPA 2003d)	This is a technical guidance and reference document for use by state, local, and tribal managers in the implementation of nonpoint source pollution management programs. It contains information on the best available, economically achievable ways to reduce pollution of surface and groundwater from agriculture. https://www.epa.gov/polluted-runoff-nonpoint-source-pollution/national-management-measures-control-nonpoint-source-0
Protecting Sources of Drinking Water: Selected Case Studies in Watershed Management (USEPA 1999)	This publication presents case studies of 17 drinking water systems that incorporated source water management and protection as an integral part of their business. The case studies focus on the lessons learned in fostering partnerships, watershed assessment, watershed land-use management, and land acquisition.
Watershed Success Stories—Applying the Principles and Spirit of the Clean Water Action Plan (USEPA 2000)	The 30 case studies included in this publication focus on restoring and maintaining water quality for surface waters that are suffering from a wide variety of issues, some of which relate to drinking water supply concerns. https://ia800203.us.archive.org/18/items/CAT11094596/CAT11094596.pdf
Drinking Water from Household Wells (USEPA 2002a)	This pamphlet provides information for homeowners on groundwater wells and how to protect their groundwater supply from contamination.
Other Guidance Material	
Agricultural Waste Management Field Handbook (USDA 2012)	This is a comprehensive guide for planning and implementing pollution control practices at a variety of agricultural operations. http://policy.nrcs.usda.gov/OpenNonWebContent.aspx?content=31473.wba

Reference Title (Citation)	Author and Description
Other Guidance Material	
Best Management Practices (BMPs) Manual—Field Guide: Monitoring, Implementation, and Effectiveness for Protection of Water Resources (Welsch et al. 2007)	This manual is one component of the US Department of Agriculture Forest Service BMP Protocol Project, which is an effort to develop a standard method for monitoring the use and effectiveness of best management practices commonly used in timber harvesting. The purpose of the BMP protocol is to create an economical, standardized, and repeatable BMP monitoring process that is completely automated, from data gathering through report generation, in order to provide measured data, ease of use, and compatibility with state BMP programs. The program includes the BMP protocol software, which is supported by this field guide and a desk reference. This field guide covers issues that need to be addressed before beginning data collection, including training field personnel and equipment specifications, as well as a description of the data input for field use of the software. http://na.fs.fed.us/pubs/misc/bmp/06/bmp_field_guide_lr.pdf
State Source Water Protection Plan Guides and Templates (ASDWA)	This document includes web links to various SWP program guides and templates from 20 different states. www.asdwa.org/index.cfm?fuseaction=page.viewpage&pageid=839
Collaboration Toolkit (Source Water Collaborative 2016)	This online guide lists the various steps of developing a collaborative group of stakeholders for advancing SWP. www.sourcewatercollaborative.org/how-to-collaborate-toolkit
Opportunities to Protect Drinking Water Sources & Advance Watershed Goals through the Clean Water Act (USEPA et al. 2014)	This state–USEPA collaboration toolkit shows how state and federal Clean Water Act (CWA) and Safe Drinking Water Act (SDWA) program staff and managers can more routinely and more intentionally coordinate CWA and SDWA activities to achieve improvements in the quality of our waters. www.asdwa.org/document/docWindow.cfm?fuseaction=document.viewDocument&documentid=3007&documentFormatId=3779
Guide for Conducting Contaminant Source Inventories for Public Drinking Water Supplies: Technical Assistance Document (USEPA, 1991a)	This technical assistance document provides information to assist state and local water managers in developing and refining methods and procedures for inventorying existing and potential sources of contamination within wellhead protection programs.
Elements of an Effective State Source Water Protection Program (ASDWA and GWPC 2008)	This document highlights various components that states can use in their SWP programs, including measurement and characterization (both statewide and at the system level); state implementation strategies; partnerships, integration, and leveraging; motivating local activity (including funding); managing and sharing information; and state regulatory programs. Case studies of successful states programs are provided as examples. www.asdwa.org/_data/n_0001/resources/live/effective%20elements%206-2008%20-%20FINAL.pdf
Information on Source Water Protection to Assist State Drinking Water Programs (ASDWA 2007)	This publication discusses implementation and information collection tools (including approaches states can use to implement SWP and that facilitate local source water data collection), coordination efforts between state drinking water programs and other state agencies, and results from a survey of state drinking water administrators regarding their perceptions of strategies for success and approaches for overcoming barriers related to CWA/SDWA integration. www.asdwa.org/_data/n_0001/resources/live/ASDWASWReportFinal21.pdf
Path to Protection—Ten Strategies for Successful Source Water Protection (Ernst and Hart 2005)	This guidance document from the Trust for Public Land summarizes findings based on experiences of five USEPA-sponsored source water demonstration pilot projects and proposes 10 strategies that will help put more state and local governments on the path to protection. Each strategy includes a case study of a state or local entity that has successfully implemented some or all of the action steps included in that strategy. http://cloud.tpl.org/pubs/water_pathtoprotection.pdf
Source Protection Handbook: Using Land Conservation to Protect Drinking Water Supplies (Hopper and Ernst 2005)	This guidance document from the Trust for Public and AWWA provides a good discussion of the reasons to use land conservation to protect drinking water supplies and a detailed discussion of the means to do so. Highlights include sections on understanding the watershed, prioritizing lands for protection, building partnerships, designing a SWP plan, financing land conservation, and protecting and managing priority lands. www.tpl.org/source-protection-handbook
Protecting the Source: Land Conservation and the Future of America's Drinking Water (Ernst 2004)	Published by the Trust for Public Land and AWWA, this report discusses the many benefits of SWP, provides case studies, and describes five key best practices that provide a framework for developing and implementing a SWP plan, including understanding the watershed; using maps and models to prioritize protection; building strong partnerships and work watershed wide; creating a comprehensive source protection plan; and developing and implementing a funding "quilt." https://www.tpl.org/sites/default/files/cloud.tpl.org/pubs/water-protecting_the_source_final.pdf.

Reference Title (Citation)	Author and Description
Other Guidance Material	
From Source to Tap: Guidance on the Multi-Barrier Approach to Safe Drinking Water (CCME 2002)	This document provides guidance on the various aspects of the multibarrier approach to safe drinking water, including SWP. It includes language and tools for drinking water systems to communicate their activities to decision makers and consumers and provides a structure for public officials to integrate health and environmental issues, collaborate and share information, and set priorities. Though tailored to a Canadian audience, much of the information is helpful for any utility or public official. www.hc-sc.gc.ca/ewh-semt/pubs/water-eau/tap-source-robinet/index-eng.php
Source Water Assessment & Protection Workshop Guide, Second Edition (Herpel 2004)	This guide provides tools to educate and motivate community members to get involved in the SWP process. It features materials for a complete informational workshop, including overheads (Microsoft PowerPoint slides), handouts, worksheets, and other materials for both the workshop presenter and participants. The material covers assessments, SWP, case studies, small group activities, implementation assistance, and evaluation. http://www.groundwater.org/action/resources.html
Getting in Step: A Guide for Conducting Watershed Outreach Campaigns, 3rd Edition (USEPA 2010)	This guide offers advice on how local governments, watershed organizations, and other stakeholders can maximize the effectiveness of public outreach campaigns to reduce nonpoint source pollution. It provides the overall framework for developing an outreach campaign plan using a step-by-step approach and gives tips on implementing the campaign plan. https://cfpub.epa.gov/npstbx/files/getnstepguide.pdf
The Practice of Watershed Protection (Schueler and Holland, 2000)	This is a comprehensive compilation of all past issues of the Center for Watershed Protection's technical journal, *Watershed Protection Techniques*. It includes articles on stormwater pollution, stormwater management practices, watershed planning, land conservation, aquatic buffers, better site design, erosion and sediment control, stream restoration, habitat and biodiversity, watershed stewardship, watershed education, and watershed monitoring. It is available for a cost at http://www.stormwatercenter.net/Library/Practice_Articles.htm.
Watershed Management for Potable Water Supply: Assessing the New York City Strategy (National Research Council 2000)	This book reports the findings of the National Research Council's (NRC's) analysis of the 1997 New York City watershed agreement's scientific validity. The NRC concludes that the watershed agreement is a good template for proactive watershed management that, if properly implemented, will maintain high water quality. The authors recommend that New York City place its highest priority on pathogenic microorganisms in the watershed and direct its resources toward improving methods for detecting pathogens, understanding pathogen fate and transport, and demonstrating that BMPs will remove pathogens. Other recommendations, which are broadly applicable to surface water supplies across the country, target buffer zones, stormwater management, water quality monitoring, and effluent trading. http://www.nap.edu/catalog.php?record_id=9677#description
Natural Infrastructure—Investing in Forested Landscapes for Source Water Protection in the United States (Gartner et al. 2013)	This guide provides a general road map for early planning and implementation steps for natural infrastructure programs. It includes a compilation of scientific knowledge, economic findings, finance mechanisms, framing language, and five case studies. www.wri.org/sites/default/files/wri13_report_4c_naturalinfrastructure_v2.pdf
Best Management Practices Databases	
International Stormwater BMP Database (Wright Water Engineers, Inc. and Geosyntec Consultants, no date)	This database provides information from more than 300 BMP studies to help improve the selection, design, and performance of different stormwater BMPs for removing a variety of pollutants. It includes performance analysis results, tools for use in BMP performance studies, monitoring guidance, and other study-related publications. www.bmpdatabase.org
Stormwater Manager's Resource Center (Center for Watershed Protection 2009)	This online information center provides technical assistance on stormwater management issues. It includes information and numerous case study models of stormwater management practices, ordinances, manuals, monitoring and assessment methods, education and public outreach programs, and other information related to programs for stormwater management, erosion and sediment control, stream buffers, and other related areas. It also includes an online library of more than 600 references. www.stormwatercenter.net
National Pollutant Removal Performance Database (version 3) (Center for Watershed Protection 2007)	This technical brief summarizes the results of more than 150 performance studies that are included in the database. It includes statistical and graphical data on removal rates for several types of green infrastructure BMPs. www.stormwaterok.net/CWP%20Documents/CWP-07%20Natl%20Pollutant%20Removal%20Perform%20Database.pdf

Reference Title (Citation)	Author and Description
Best Management Practices Databases	
Performance of Green Infrastructure, Databases and Summary Reports (USEPA 2016)	This USEPA web page provides links to various databases and summary reports that include statistics on performance efficiency for the ability of different stormwater controls to reduce pollutants. https://www.epa.gov/green-infrastructure/performance-green-infrastructure
Efficiency of Urban Stormwater Best Management Practices: A Literature Review (Hallock 2007)	This is a brief but useful summary of expected pollutant concentrations for various land uses, and efficiency of different types of BMPs. It also includes an annotated bibliography of various references and databases that report on BMP effectiveness. https://fortress.wa.gov/ecy/publications/summarypages/0703009.html
Information on Pollutant Removal by BMPs (Minnesota Pollution Control Agency, no date)	This web-based stormwater manual contains information on pollutant removal for various pollutants. There are numerous links to additional information on pollutant removal. http://stormwater.pca.state.mn.us/index.php/Information_on_pollutant_removal_by_BMPs
Water Research Foundation Project (WRF) Reports: www.waterresearchfoundation.org	
Knowledge Portals: Source Water Protection and Management (Water Research Foundation)	This document includes a series of fact sheets and lists of related WRF projects related to algae and cyanobacteria; lake and reservoir management; land use and water quality; mussels, milfoil, and other nuisance organisms; and extreme weather events and impacts. http://www.waterrf.org/knowledge/source-water-protection-and-management/Pages/default.aspx
A Methodology for Locating and Managing Dynamic Potential Source Water Contaminant Data (project 4581; Corona Environmental Consulting LLC, in progress 2016)	This project involves development of a detailed methodology for how information about the storage of chemicals upstream of drinking water intakes can be located, extracted, quality controlled, organized, and maintained by a water system. It also includes development and pilot testing of a software tool (WaterSuite) that provides utilities with access to information necessary for mitigation and response planning in anticipation of potential acute risks to source water, including releases from chemical storage tanks, pipelines, discharges, transportation accidents, oil and gas wells. WaterSuite's Source Water Protection Mapping Tool aggregates large volumes of federal, state, local, private, and proprietary data to help prioritize SWP efforts and emergency response capabilities. It includes a platform for Cloud-based data submission, as well as storing, viewing, and interpreting historic and real-time water quality monitoring data, and it accepts both automated and grab sample data. www.watersuite.com
Source Water Protection Vision and Roadmap (project 4176; Sklenar et al. 2012)	The overall goal of this project was to develop a vision and road map that can guide US water utilities and supporting groups with a unified strategy for coherent, consistent, cost-effective, and socially acceptable source water protection programs. The following two approaches were used to achieve this: first, a summary of the state of source water protection in the United States was developed, and then a discussion among representatives of the water industry and related stakeholders was held, and a consensus regarding a common vision and unified strategy was developed. Combined, these efforts were used to create a supporting vision and road map for source water protection that, along with supporting tools, can be used to help motivate, catalyze, and plan new SWP programs as well as improve existing programs. Common obstacles to SWP are discussed and means to overcome those obstacles are presented. Thirteen water utility SWP case studies are also provided, as is a literature review that includes a history of SWP laws in the United States.
Source Water Protection Cost/Benefit Tool (project 4143; Oxenford et al. 2010)	This project includes a compilation of information on the triple-bottom line costs and benefits for various best-management practices and organizational approaches for SWP. The project's output is a web-based tool for accessing that information. www.swptool.org/index.cfm
Drinking Water Source Protection Through Effective Use of TMDL Processes (project 4007, Sklenar and Blake 2010)	Researchers for this project investigated successful strategies used by drinking water utilities to protect source waters using the total maximum daily loads (TMDL) regulatory process and described and evaluated specific measures that have been used to include drinking water objectives in TMDLs.
Mitigating Impacts of Changes in Watershed Vegetation on Source Water Quality and Quantity (project 4009; Strange et al. 2009)	Researchers for this project investigated the impacts of short-term, catastrophic or longer-term natural and human-caused changes to vegetative cover on the quality and quantity of source waters. It includes discussion of prevention and mitigation response strategies undertaken by utilities. Eight watershed events were investigated. Natural events included wildfire; storm events such as hurricanes, tornadoes, and floods; mudslides; insect pests and pathogens; and drought. Human-caused events included introduction of invasive species, timber harvesting, and land conversion such as agricultural development and urbanization.

Reference Title (Citation)	Author and Description
Water Research Foundation Project (WRF) Reports: www.waterresearchfoundation.org	
Source Water Protection for Concentrated Animal Feeding Operations: A Guide for Drinking Water Utilities (project 3020; Gullick et al. 2006)	This report presents information and guidance regarding the basics of concentrated animal feeding operations and related agricultural activities; characteristics and quantity of the wastes produced; the fate and transport of these contaminants in the environment; potential contaminant impacts on source water quality; and strategies that water utilities and animal feeding operations can use to control and monitor the release of these contaminants to drinking water sources. An electronic information center is also available that includes much of the report information. http://www.eetinc.com/tools/cafo/afo.html
Development of Event-Based Pathogen Monitoring Strategies for Watersheds (project 2671; Sturdevant Rees et al., 2006)	For this report, researchers used sampling, analytical, and statistical methods to advance current understanding of the variability of pathogen occurrence and transport through watersheds and to develop methodologies aimed at minimizing risks of waterborne disease outbreaks in drinking water supplies. Specifically, the research objectives were to develop and validate a strategy for selection of sampling locations, frequencies, and methods to accurately depict pathogen occurrence in relation to various sources within watersheds during and after weather, hydrologic, or land-use events.
Water Utility/Agricultural Alliances: Working Together for Cleaner Water (project 2781, Fletcher et al. 2005)	This report identifies strategies for drinking water utilities to build successful SWP alliances with agricultural producers at the local, state, and national level. The project was designed to provide drinking water utilities with guidance on how to build alliances with farmers and agricultural organizations in order to promote agricultural practices that minimize runoff and protect the quality of drinking water sources. Guidance was obtained in part from 20 case studies and an expert workshop.
Transport of Surface Water Pathogens in Watersheds (project 2684, Davies et al. 2005)	This report characterizes the physicochemical interactions of key pathogen groups (e.g., *Cryptosporidium*, enteric viruses, and bacteria) associated with sediment and organic matter in watersheds. It also identifies and quantifies pathogen sources, fate, and hydrologic transport under various environmental conditions. The project's principal goal was to facilitate the development of predictive models to describe expected concentrations of waterborne pathogens at critical downstream locations. The project report includes a CD-ROM that contains raw data and photos.
Total Maximum Daily Loads (TMDLs) and Drinking Water Utilities (project 2944; Rosen et al., 2005)	This report includes results of a workshop of experts that was held to identify gaps in the current knowledge base and potential research needs regarding TMDLs. It also identifies the benefits and impacts to drinking water utilities from the TMDL program and was intended to contribute to opening and expanding lines of communication between TMDL regulatory agencies and drinking water utilities.
Performance and Whole-Life Costs of Best Management Practices and Sustainable Urban Drainage Systems (project 2880, Lampe et al. 2004 and 2005)	This report provides an understanding of the long-term costs and environmental impacts of sustainable urban drainage systems (SUDS) and BMPs. Phase 1 of the project includes a literature review and a survey of stormwater authorities and organizations in the United States and the United Kingdom to identify the most commonly used BMPs and SUDS and to determine the availability of data on their cost and performance. It also describes the performance of different BMPs and SUDS, long-term maintenance needs, and the impact of maintenance activities on performance. Furthermore, it discusses establishment of protocols for whole-life costs and performance data for BMPs and SUDS. In phase 2 the operation of selected BMPs and SUDS was monitored for 1 year in terms of pollutant removal and hydrologic/hydraulic efficiency, as well as applicability of their design criteria and maintenance regimens.
Demonstrating Benefits of Wellhead Protection Programs (project 2778, Williams and Fenske 2004)	The objectives of this project were to identify the key elements of wellhead protection programs; identify the costs to develop and implement local wellhead protection programs; compile a list of benefits of wellhead protection programs (including water quality, economics, and ecological and other non-monetary benefits); and develop a generic method to measure and quantify those benefits.
Chemical Occurrence Data Sets for Source Water Assessments (project 2756; Stevens et al. 2003)	This report identifies, lists, and characterizes chemical occurrence data sets that can be used in conducting source water assessments. The report provides a directory of relevant source water assessment data to serve as a resource for agencies, water utilities, consultants, and other stakeholders in improving source water assessment regulations and programs. Twelve major national databases, 9 regional databases, and 71 state data sets were inventoried and reviewed based on a variety of descriptors (metadata). A companion website was developed but unfortunately is no longer available.

Reference Title (Citation)	Author and Description
Water Research Foundation Project (WRF) Reports: www.waterresearchfoundation.org	
Impacts of Major Point and Non-Point Sources on Raw Water Treatability (project 2616, Pyke et al. 2003)	Researchers evaluated the effects of major point and nonpoint pollutant sources on raw water quality, drinking water treatability, and water treatment costs. They also evaluated the potential for using agricultural and urban BMPs to mitigate these effects and compared the efficacy and cost of BMPs in the watershed with increased treatment at the water treatment plant (WTP). It includes a literature review that describes the extent and impact of agricultural nonpoint source pollution; discussion of pollutant removal mechanisms and performance of common agricultural BMPs; an overview of watershed modeling that describes watershed models, their limitations, and the resources needed to use them; an overview of WTP modeling including discussion of underlying model assumptions; results of watershed model simulations that show the relative contributions of different point and nonpoint sources and the effects of BMP implementation on pollutant loads of solids, phosphorus, and total organic carbon to receiving waters; and results of WTP model simulations that show the impact of changes in pollutant loads on the operations, costs, and effluent quality of four types of WTPs.
Source Water Protection Reference Manual (project 2651; CDM 2002)	Researchers created a CD-ROM for water utilities that provides guidance for developing a comprehensive watershed management plan, as well as descriptions of regulations and policy related to watershed management for all 50 states, the US federal government, the United Kingdom, and Canada. It includes case study information from 85 water suppliers about source water characterization, management plans, BMP implementation guidance, and comparison of BMP effectiveness and includes a search engine and web hot links.
Guidance to Utilities on Building Alliances with Watershed Stakeholders (project 468, Raucher and Goldstein, 2001)	This report discusses the need for and procedures for building win–win alliances between water utilities and various stakeholders for the purpose of overcoming constraints to planning, managing, and developing long-term sustainable drinking water supplies. The report identifies typical watershed stakeholders and their objectives in basin planning. Guidance is given in the form of examples, worksheets, summaries, reference lists, and other tools. It is designed for both experienced and inexperienced utilities at any stage in an alliance process.
Design of Early Warning and Predictive Source-Water Monitoring Systems (project 2527; Grayman et al. 2001)	For this this project, researchers assessed early warning systems (EWSs), identified existing and emerging monitoring options, and developed guidelines for design and operation of an EWS. In addition, a general purpose, one-dimensional riverine spill model was developed and applied to the Ohio River; a systematic method for designing and operating EWSs was identified; and a risk-based model using Monte Carlo simulation techniques was developed and demonstrated for design of EWSs. The research included a literature review, utility survey, site visits, case studies of advanced EWSs around the world, and development of the modeling aspects of the study.
Effective Watershed Management for Surface Water Supplies (project 317, Robbins et al. 1991)	This report provides guidance for water utility managers and local governments to develop effective watershed protection programs for their surface water supplies. It emphasizes practical, effective solutions and techniques to apply to agricultural land, forest management activities, and urban areas. Much of the information is based on a national survey of water utilities and state regulatory agencies and 24 case studies of successful watershed protection programs across the country conducted as part of the project.

Notes: Most USEPA materials are available from the National Service Center for Environmental Publications (NSCEP) at:
   US Environmental Protection Agency
   National Service Center for Environmental Publications (NSCEP)
   P.O. Box 42419; Cincinnati, OH 45242-0419
   E-mail: nscep_nepis.tech@epa.gov; website: www.epa.gov/nscep

SECTION 9: REFERENCES

ASDWA (Association of State Drinking Water Administrators). 2007. Information on Source Water Protection to Assist State Drinking Water Programs. Arlington, Va.: ASDWA. www.asdwa.org/_data/n_0001/resources/live/ASDWASWReportFinal21.pdf.

———. 2008. Summary of State Source Water Protection Survey Responses. Presented at the 2008 ASDWA/GWPC State Source Water Protection Workshop, Colorado Springs, Colo., October 23, 2008. Arlington, Va.: ASDWA.

ASDWA and GWPC (Association of State Drinking Water Administrators, and Ground Water Protection Council). 2008. *Elements of an Effective State Source Water Protection Program* (Second Version). Arlington, Va.: ASDWA. www.asdwa.org/_data/n_0001/resources/live/effective%20elements%206 -2008%20-%20FINAL.pdf.

AWWA (American Water Works Association). 1999. *Source Water Protection: Effective Tools and Techniques You Can Use* (1999 Participant Manual). Denver, Colo.: AWWA.

———. 2006. *Source Water Protection Best Management Practices* (2006 Participant Manual). Denver, Colo.: AWWA.

———. 2007. Manual of Water Supply Practices. *M50, Water Resources Planning.* 2nd ed. Chapter 11: Watershed Management and Groundwater Protection. Denver, Colo.: AWWA. http://www.awwa.org/store/productdetail.aspx ?productid=6738.

———. 2010. *Policy Statement on Quality of Public Water Supply Sources* (last updated January 17, 2010). Denver, Colo.: AWWA. http://www.awwa.org /about-us/policy-statements/policy-statement/articleid/210/quality-of- water-supply-sources.aspx.

———. 2014. ANSI/AWWA G300, Source Water Protection. Denver, Colo.: AWWA. http://www.awwa.org/store/productdetail.aspx?productId=39814230.

CCME (Canadian Council of Ministers of the Environment). 2002. *From Source to Tap: Guidance on the Multi-Barrier Approach to Safe Drinking Water.* Developed by the Federal-Provincial-Territorial Committee on Drinking Water of the Federal-Provincial-Territorial Committee on Environmental and Occupational Health and the Water Quality Task Group of the CCME. Winnipeg, Manitoba, Canada: CCME. http://www.hc-sc.gc.ca/ewh-semt /pubs/water-eau/tap-source-robinet/index-eng.php.

CDM (Camp Dresser & McKee, Inc.). 2002. *Source Water Protection Reference Manual* (CD-ROM). Denver, Colo.: Water Research Foundation and AWWA.

Center for Watershed Protection. 2000. Stormwater Manager's Resource Center (website). Ellicott City, Md.: Center for Watershed Protection. www.storm watercenter.net.

Center for Watershed Protection (SWP). 2007. National Pollutant Removal Performance Database—Version 3. Ellicott City, Md.: Center for Watershed Protection. Available: www.stormwaterok.net/CWP%20Documents/CWP-07%20Natl%20Pollutant%20Removal%20Perform%20Database.pdf.

Central Arkansas Water. 2007. *Lake Maumelle Watershed Management Plan.* Prepared by Tetra Tech, Inc. http://www.carkw.com/wp-content/uploads/2011/09/Lake_Maumelle_Watershed_Mgmt_Plan_May_07_reduced.pdf.

Davies, Cheryl, Christine Kaucner, Nanda Altavilla, Nicholas Ashbolt, Christobel Ferguson, Martin Krogh, Wim Hijnen, Gertjan Medema, and Daniel Deere. 2005. *Fate and Transport of Surface Water Pathogens in Watersheds.* Denver, Colo.: Water Research Foundation.

Davis, Chris (ed.). 2008. *Watershed Management for Drinking Water Protection.* Denver, Colo.: AWWA; St. Leonards, NSW, Australia: Australian Water Association.

Edge, Tom A., and Karl A. Schaefer (eds.). 2006. *Microbial Source Tracking in Aquatic Ecosystems: The State of Science and an Assessment of Needs.* NWRI Scientific Assessment Report Series No. 7. Burlington, Ontario: National Water Research Institute.

Ernst, Caryn. 2004. *Protecting the Source: Land Conservation and the Future of America's Drinking Water.* Washington, D.C.: Trust for Public Land.

Ernst, Caryn, and Kelley Hart. 2005. *Path to Protection: Ten Strategies for Successful Source Water Protection.* Washington, D.C.: Trust for Public Land. http://cloud.tpl.org/pubs/water_pathtoprotection.pdf.

Fletcher, Angie, Susan Davis, and Grantley Pyke. 2005. *Water Utility/Agricultural Alliances: Working Together for Cleaner Water.* Denver, Colo.: Water Research Foundation.

Fletcher, Angie, Susan Davis, and Grantley Pyke. 2005. *Water Utility/Agricultural Alliances: Working Together for Cleaner Water.* Denver, Colo.: Water Research Foundation.

Gartner, Todd, James Mulligan, Rowan Schmidt, and John Gunn (eds.). 2013. Natural Infrastructure – Investing in Forested Landscapes for Source Water Protection in the United States. Washington, DC: World Resources Institute. https://www.wri.org/sites/default/files/wri13_report_4c_naturalinfrastructure_v2.pdf

Grayman, Walter M., Rolf A. Deininger, and Richard M. Males. 2001. *Design of Early Warning and Predictive Source-Water Monitoring Systems.* Denver, Colo.: Water Research Foundation.

Gullick, Richard W. 2014. Source Water Protection: Perspectives of the Past, Present, and Future. *Jour. AWWA*, 106(8):164-174.

———. 2003. Committee Connection: AWWA's Source Water Protection Committee Outlines How to Maintain the Highest Quality Source Water. *Jour. AWWA*, 95(11):36–42.

Gullick, Richard W., Richard A. Brown, and David A. Cornwell. 2006. *Source Water Protection for Concentrated Animal Feeding Operations: A Guide for Drinking Water Utilities*. Denver, Colo.: Water Research Foundation.

Hallock, David. 2007. Efficiency of Urban Stormwater Best Management Practices: A Literature Review. Olympia, Washington: Washington State Department of Ecology, Publication No. 07-03-009. http://lshs.tamu.edu/docs/lshs/end-notes/efficiency%20of%20urban%20stormwter%20bmps_a%20literature%20review-0674052123/efficiency%20of%20urban%20stormwter%20bmps_a%20literature%20review.pdf.

Herpel, Rachael. 2004. *Source Water Assessment and Protection Workshop Guide. 2nd ed.* Lincoln, Neb.: Groundwater Foundation. www.groundwater.org/gi/swap.html.

Hopper, Kim, and Caryn Ernst. 2005. *Source Protection Handbook: Using Land Conservation to Protect Drinking Water Supplies*. Washington, D.C.: Trust for Public Land and Denver, Colo.: AWWA. www.tpl.org/source-protection-handbook.

Lampe, Les, Howard Andrews, Michael Barrett, Bridget Woods-Ballard, Peter Martin, Charles Glass, Neil Weinstein, and Chris Jeffries. 2004. *Post-Project Monitoring of BMPs/SUDs to Determine Performance and Whole Life Costs* (Report for Phase 1). Alexandria, Va.: Water Environment Research Foundation.

Lampe, Les, Howard Andrews, Michael Barrett, Bridget Woods-Ballard, Richard Kellagher, Peter Martin, Chris Jeffries, and Matt Hollon. 2005. *Performance and Whole-Life Costs of Best Management Practices and Sustainable Urban Drainage Systems* (Final Report for Phase 2). Alexandria, Va.: Water Environment Research Foundation.

LDEQ (Louisiana Department of Environmental Quality). nd-a. Final Potential Susceptibility Analysis of a Ground Water Source of Public Drinking Water. www.deq.louisiana.gov/portal/Portals/0/evaluation/aeps/swap/Final_Susceptibility_Analysis_Form_GW_Systems.pdf .

———. nd-b. Final Potential Susceptibility Analysis of a Surface Water Source of Public Drinking Water. www.deq.louisiana.gov/portal/Portals/0/evaluation/aeps/swap/Final_Susceptibility_Analysis_Form_SW_Systems.pdf.

Long, S.C., and J.D. Plummer. 2004. Assessing Land Use Impacts on Water Quality Using Microbial Source Tracking. *Jour. AWRA*, 40(6):1433–1438.

Minnesota Pollution Control Agency. No date. *Information on Pollutant Removal by BMPs* (website). http://stormwater.pca.state.mn.us/index.php/Information_on_pollutant_removal_by_BMPs.

National Research Council, Committee to Review the New York City Watershed Management Strategy, 2000. *Watershed Management for Potable Water Supply: Assessing the New York City Strategy*. Washington, D.C.: National Academy Press. www.nap.edu/catalog.php?record_id=9677#description.

NEIWPCC (New England Interstate Water Pollution Control Commission). 2000. Source Protection: A National Guidance Manual for Surface Water Supplies. Lowell, Mass.: NEIWPCC.

O'Connor, D.R., 2002. Part Two Report of the Walkerton Inquiry: A Strategy for Safe Drinking Water. Ontario Ministry of the Attorney General. Publications Ontario, Toronto, Ontario, Canada. http://www.archives.gov.on.ca/en/e_records/walkerton/report2/index.html.

Oxenford, Jeff, et al. Source Water Protection Cost/Benefit Tool (website: http://www.swptool.org/index.cfm). Denver, Colo.: Water Research Foundation.

PWD (Philadelphia Water Department). 2002a. *Belmont and Queen Lane Treatment Plants* (PWSID #1510001). Source Water Assessment Report. Philadelphia, Pa.: PWD. http://www.phillywatersheds.org/doc/Schuylkill_SWA_lowres.pdf.

———. 2002b. *Baxter Water Treatment Plant* (PWSID #1510001). Source Water Assessment Report. Philadelphia, Pa.: PWD. http://www.phillywatersheds.org/doc/Delaware_SWA.pdf.

———. 2006. *The Schuylkill River Watershed Source Water Protection Plan (Belmont and Queen Lane Surface Water Intakes)*. Philadelphia, Pa.: PWD. http://www.phillywatersheds.org/doc/Schuylkill_SWPP_2006.pdf.

———. 2007. *The Delaware River Watershed Source Water Protection Plan (Baxter Water Treatment Plant Surface Water Intake)*. Philadelphia, Pa.: PWD. http://www.phillywatersheds.org/doc/Delaware_SWPP_2007.pdf.

Plummer, J.D., and S.C. Long. 2007. Monitoring Source Water for Microbial Contamination: Evaluation of Water Quality Measures. *Water Research,* 41(16):3716–3728.

Pyke, Grantley W., William C. Becker, Richard Head, and Charles R. O'Melia. 2003. *Impacts of Major Point and Non-Point Sources on Raw Water Treatability.* Denver, Colo.: Water Research Foundation.

Raucher, Robert S., and James Goldstein. 2001. *Guidance to Utilities on Building Alliances with Watershed Stakeholders.* Denver, Colo.: Water Research Foundation.

Robbins, R.W., J.L. Glicker, D.M. Bloem, and B.M. Niss. 1991. *Effective Watershed Management for Surface Water Supplies.* Denver, Colo.: Water Research Foundation.

Rosen, Jeffrey S., Jose A.H. Sobrinho, Paul L. Freedman, and Wendy M. Larson. 2005. *Total Maximum Daily Loads (TMDLs) and Drinking Water Utilities.* WRF report 91049F. Denver, Colo.: Water Research Foundation.

Schueler, Thomas R., and Heather K. Holland (eds.). 2000. *The Practice of Watershed Protection.* Ellicott City, Md.: Center for Watershed Protection.

Sham, Chi Ho, Richard W. Gullick, Sharon C. Long, and Pamela P. Kenel (2010). *Source Water Protection: Operational Guide to AWWA Standard G300.* Denver, Colo.: AWWA.

Sklenar, Karen, Chi Ho Sham, and Richard W. Gullick. 2012. *Source Water Protection Vision and Roadmap.* Denver, Colo.: Water Research Foundation.

Sklenar, Karen and Laura J. Blake. 2010. *Drinking Water Source Protection Through Effective Use of TMDL Processes.* Denver, Colo.: Water Research Foundation.

Source Water Collaborative. 2016. Collaboration Toolkit (online guide). www.sourcewatercollaborative.org/how-to-collaborate-toolkit.

Stevens, Krystin B., Jose A.H. Sobrinho, Jeffrey S. Rosen, and Christopher Crockett. 2003. *Chemical Occurrence Data Sets for Source Water Assessments.* Denver, Colo.: Water Research Foundation.

Strange, Elizabeth M., Diana R. Lane, and Charles N. Herrick. 2009. *Utility Guidance for Mitigating Catastrophic Vegetation Change in Watersheds.* Denver, Colo.: Water Research Foundation.

Sturdevant Rees, Paula L., Sharon C. Long, Rebecca Baker, Daniel H. Bordeau, Routing Pei, and Paul K. Barten. 2006. *Development of Event-Based Pathogen Monitoring Strategies for Watersheds.* Denver, Colo.: Water Research Foundation.

Tiemann, Mary. 2008. *CRS Report for Congress—Safe Drinking Water Act: A Summary of the Act and Its Major Requirements.* Order Code RL31243; updated May 21, 2008. Washington, D.C.: Congressional Research Service. www.ncseonline.org/NLE/CRSreports/08Jun/RL31243.pdf.

USDA (US Department of Agriculture). 1999. *Agricultural Waste Management Field Handbook* (National Engineering Handbook, Part 651). USDA Natural Resources Conservation Service. 210-VI, NEH-651. First published in 1992; last updated 2012. http://policy.nrcs.usda.gov/viewerFS.aspx?hid=21430.

USEPA (US Environmental Protection Agency). 1987. *Guideline for Delineation of Wellhead Protection Areas.* EPA 440-6-87/010. Washington, D.C.: USEPA, Office of Ground-Water Protection.

———. 1989. *Wellhead Protection Programs: Tools for Local Governments.* EPA 440-6-89-002. Washington, D.C.: USEPA, Office of Water.

———. 1991a. *Guide for Conducting Contaminant Source Inventories for Public Drinking Water Supplies: Technical Assistance Document.* EPA 570-9-91-014. Washington, D.C.: USEPA.

———. 1991b. *Managing Ground Water Contamination Sources in Wellhead Protection Areas: A Priority Setting Approach.* EPA 570-9-91-023. Washington, D.C.: USEPA, Office of Ground Water and Drinking Water.

———. 1993. *Guidance Specifying Management Measures for Sources of Nonpoint Pollution in Coastal Waters.* EPA 840-B-92-002. Washington, D.C.: USEPA, Office of Water. www.epa.gov/polluted-runoff-nonpoint-source-pollution/guidance-specifying-management-measures-sources-nonpoint.

———. 1997a. *Delineation of Source Water Protection Areas: A Discussion for Managers.* Part 1: A Conjunctive Approach for Ground Water and Surface Water. EPA 816-R-97-012. Washington, D.C.: USEPA, Office of Water.

———. 1997b. *Guidelines for Wellhead and Springhead Protection Area Delineation in Carbonate Rocks.* EPA 904-B-97-003. Atlanta, Ga.: USEPA, Ground Water Protection Branch, Region 4.

———. 1997c. *State Source Water Assessment and Protection Programs: Final Guidance.* EPA 816-R-97-009. Washington, D.C.: USEPA Office of Water.

———. 1999. *Protecting Sources of Drinking Water: Selected Case Studies in Watershed Management.* EPA 816-R-98-019, April 1999. Washington, D.C.: USEPA Office of Water.

———. 2000. *Watershed Success Stories—Applying the Principles and Spirit of the Clean Water Action Plan.* Washington, D.C.: USEPA.

————. 2001a. *National Source Water Contamination Prevention Strategy: Seventh Draft for Discussion* (April 2001). Washington, D.C.: USEPA Office of Ground Water and Drinking Water. http://permanent.access.gpo.gov/lps21800/www .epa.gov/safewater/protect/strateg7.pdf; a companion discussion of national SWP information needs is also available at http://permanent.access.gpo.gov /lps21800/www.epa.gov/safewater/protect/1205meas.pdf.

————. 2001b. *Protecting and Restoring America's Watersheds: Status, Trends, and Initiatives in Watershed Management.* EPA 840-R-00-001. Washington, D.C.: USEPA Office of Water.

————. 2002a. *Drinking Water from Household Wells.* Washington, D.C.: USEPA Office of Ground Water and Drinking Water.

————. 2002b. *Source Water Protection: Best Management Practices and Other Measures for Protecting Drinking Water Supplies.* Washington, D.C.: USEPA Office of Ground Water and Drinking Water. https://cfpub.epa.gov/ watertrain/pdf/swpbmp.pdf.

————. 2002c. *Source Water Protection Practices Bulletins.* Washington, D.C.: USEPA Office of Ground Water and Drinking Water. http://cfpub.epa.gov /safewater/sourcewater/sourcewater.cfm?action=Publications&view=filter& document_type_id=103.

————. 2003a. *Annotated Bibliography of Source Water Protection Materials.* EPA 816-F-03-010, June 2003. Washington, D.C.: USEPA Office of Ground Water and Drinking Water. www.sourcewaterpa.org/wp-content/uploads/2010/10/ Annotated-Bibliography-of-Source-Water-Protection-Materials.pdf.

————. 2003b. *Drinking Water Inspector's Field Reference: For Use When Conducting a Sanitary Survey of a Small Ground Water System.* EPA 816-R-03-023. Washington, D.C.: USEPA, Office of Water.

————. 2003c. *Drinking Water Inspector's Field Reference: For Use When Conducting a Sanitary Survey of a Small Surface Water System.* EPA 816-R-03-022. Washington, D.C.: USEPA, Office of Water.

————. 2003d. *National Management Measures to Control Nonpoint Source Pollution from Agriculture.* EPA 841-B-03-004. Washington, D.C.: USEPA Office of Water. www.epa.gov/polluted-runoff-nonpoint-source-pollution/ national-management-measures-control-nonpoint-source-0.

———. 2005. *Microbial Source Tracking Guide Document.* EPA/600-R-05-064. Cincinnati, Ohio: USEPA Office of Research and Development. http://twri.tamu.edu/docs/bacteria-tmdl/epa%20mstguide%206-05.pdf.

———. 2006. *How-to Manual: Update and Enhance Your Local Source Water Protection Assessments.* Washington, D.C.: USEPA Office of Ground Water and Drinking Water. https://www.epa.gov/sites/production/files/2015-06/documents/816k06004.pdf.

———. 2008. *Handbook for Developing Watershed Plans to Restore and Protect Our Waters.* EPA 841-B-08-002. Washington, D.C.: USEPA Office of Water. www.epa.gov/polluted-runoff-nonpoint-source-pollution/handbook-developing-watershed-plans-restore-and-protect.

———. 2010. *Getting in Step: A Guide for Conducting Watershed Outreach Campaigns, 3rd Edition.* EPA 841-B-10-002, 163+ pp. Washington, D.C.: USEPA. https://cfpub.epa.gov/npstbx/files/getnstepguide.pdf.

———. 2013. Federal Funding Opportunities for Source Water Protection. EPA 816-K-13-001. https://www.epa.gov/sites/production/files/2015-04/documents/epa816k13001.pdf.

———. 2016. *Performance of Green Infrastructure, Databases and Summary Reports* (website) Washington, D.C.: USEPA. https://www.epa.gov/green-infrastructure/performance-green-infrastructure.

USEPA, ACWA, ASDWA, and GWPC. 2014. Opportunities to Protect Drinking Water Sources and Advance Watershed Goals through the Clean Water Act. Washington, D.C.: USEPA. www.asdwa.org/document/docWindow.cfm?fuseaction=document.viewDocument&documentid=3007&documentFormatId=3779.

Williams, Mark B., and Bruce A. Fenske. 2004. *Demonstrating Benefits of Wellhead Protection Programs.* Denver, Colo.: Water Research Foundation.

Wright Water Engineers, Inc., and Geosyntec Consultants, Inc. 2010. International Stormwater BMP Database. www.bmpdatabase.org.

Welsch, David, Roger Ryder, and Tim Post. 2007. Management Practices (BMPs) Manual – Field Guide: Monitoring, Implementation, and Effectiveness for Protection of Water Resources. NA–FR–02–06. Newtown Square, PA: USDA Forest Service. http://na.fs.fed.us/pubs/misc/bmp/06/bmp_field_guide_lr.pdf.

SECTION 10: WORKSHEETS

The following worksheets can be used during the source water protection (SWP) evaluation process to help ensure that the most relevant factors have been addressed. A review of these worksheets could help to identify subject areas that have not yet been considered.

There are six worksheets that cover each of the six basic steps listed in ANSI/ AWWA Standard. Additional worksheets provide further detail for some of the steps. Please note that there may be some repetition for select factors, especially between the main worksheet for one of the six steps and the subsequent more detailed worksheets for that same subject area.

Criteria are included for rating the relative success of the approach for individual components within the worksheets. These criteria can be used to help determine the extent of conformance with ANSI/AWWA Standard, as well as the overall success of the planning and implementation process. However, not all aspect of these worksheets need to be covered for a SWP program to be successful, and some aspects may not be relevant to certain situations/drinking water sources.

- Worksheet A: Developing a Vision and Coordinating with Stakeholders
- Worksheet B: Source Water and Source Water Protection Area Characterization
- Worksheet B-1: Delineating the Source Area of Concern
- Worksheet B-2: Water Quality Information
- Worksheet B-3: Contaminant Source Inventory Data
- Worksheet B-4: Land Use Analysis
- Worksheet B-5: Physical Barrier Effectiveness Determination
- Worksheet B-6: Intake Structure
- Worksheet B-7: Filling Information Gaps and Needs
- Worksheet B-8: Analysis of Vulnerability/Susceptibility
- Worksheet C: Source Water Protection Goals
- Worksheet D: Source Water Protection Action Plans
- Worksheet D-1: Prioritization and Planning
- Worksheet D-2: Contingency Planning
- Worksheet E: Program Implementation
- Worksheet E-1: Assessing Key Milestones
- Worksheet E-2: Roles and Responsibilities

- Worksheet E-3: Resources
- Worksheet E-4: Water Quality Monitoring
- Worksheet E-5: Biological/Habitat Monitoring
- Worksheet E-6: Stakeholder and Public Relations
- Worksheet F: Program Evaluation and Revision
- Worksheet G: Verification and Record Keeping

Worksheet A: Developing a Vision and Coordinating with Stakeholders

Rationale/Goal

Use this worksheet to assess progress toward developing a vision for the SWP program and for involving key stakeholders throughout the procesws. The vision is a statement of the utility's commitment to SWP and helps align priorities and resources. Partnering with stakeholders is typically essential for a successful SWP program. This can help to jointly identify a variety of opportunities and alternatives and bring various resources to the program effort, such as expertise, labor, and funding.

Assessment

	High	Medium	Low	None
Vision				
1. Is there a utility vision, mission statement, or policy that specifically addresses SWP?	Yes			No
2. Has the vision, mission statement, or policy been adopted by the utility governing board?	Yes, direct approval	Indirect or tacit acceptance		No
3. Is the SWP vision, mission statement, or policy distributed and understood throughout the organization?	All staff	Management level only	Board of directors only	No (known only by the developers)
4. Does the mission statement recognize that SWP is one of the multiple barriers for drinking water production?	Yes			No
5. Does the policy or utility mission statement include commitment of, or intention to commit, sufficient resources?	Yes			No
6. Is there a process for regular or periodic review of the SWP vision? (and when was the SWP vision last reviewed?)	Yes (date of last review)			No
7. Is the SWP vision available to the public (in a consumer confidence report, annual report, other outreach materials, and/or the utility's website)?	Yes			No
Stakeholders				
8. Have key stakeholders been identified and involved in development of the mission statement (e.g., did outside entities have an opportunity to comment)?	Extensive stakeholder involvement	Partial survey and contact	Minimal effort	No stakeholder involvement
9. Have key stakeholders been identified and involved in SWP program planning?	Extensive stakeholder involvement	Partial involvement	Minimal involvement	No stakeholder involvement
10. Have the existing SWP programs and initiatives of the key stakeholders and other organizations been identified (including water quality monitoring, best management practices, land conservation, education, and similar activities)?	Extensive identification and compilation of information	Partial identification and compilation of information	Minimal identification and compilation of information	None
11. Have key stakeholders been identified and involved in SWP program implementation?	Extensive stakeholder involvement	Partial involvement	Minimal involvement	No stakeholder involvement

Worksheet B: Source Water and Source Water Protection Area Characterization

Rationale/Goal

Use this worksheet to learn the main, basic components of the characterization process. A more detailed evaluation is provided in Worksheets B-1 through B-8.

Assessment

Assessment	High	Medium	Low	None
1. Have the SWP area(s) and area(s) of concern been delineated?	Complete	Partial	Just started	No
2. Do water quality data exist for the source water at intakes or wells?	Extensive and comprehensive sets of data (more than 10 years worth)	Extensive but not comprehensive (e.g., missing certain water quality parameters and data records)	Partial sets of data	No data
3. Do inventories, records, or knowledge of actual and potential contaminant sources and associated land-use information exist?	GIS maps with all sources and uses	Partial mapping of sources and uses	Partial listing of source and uses	No data
4. Is the information in a useable format?	GIS data layers available	Computer files available	Paper files available	No data or unusable data format
5. Have existing management activities and pollution control practices in the SWP area been identified and evaluated?	Complete census and analyses	Partial survey and analyses	Limited coverage	No evaluation
6. Have source water area stakeholders and land owners, their roles, and their initiatives been identified? Are working relationships with these groups established?	Extensive identification and contact efforts	Partial survey and contact	Minimal effort	No identification, and no stakeholder involvement
7. Has a source water susceptibility analysis been conducted?	Yes, complete analysis	Yes, partial analysis	Limited analysis	No
8. Are relevant personnel aware of applicable federal/state/provincial/ local regulations?	Yes, and have comprehensive knowledge	Yes, but not sure about some of the regulations (e.g., other non-SDWA regulations)	Yes, but not sure about most of the regulations involved	No knowledge of regulations
9. Is there a process for periodic updating of the source protection area characterization?	Yes	Process is under development		No

Worksheet B-1: Delineating the Source Area of Concern

Rationale/Goal

Use this worksheet to determine the physical setting of the water supply and the area of contribution. This provides a basis for focusing data compilation, evaluation, and protection activities. Answering the following questions will help determine the quality of the source water area inventory map. If the sources are groundwater under the direct influence of surface water, then check both sets of criteria.

Assessment

Refer to questions 1–6 if assessing a surface water source, and questions 7–13 for groundwater sources.

	High	Medium	Low
Surface water supply			
1. Is there a topographic map(s) of your water supply area available?	Yes, electronic/GIS maps of topography that are detailed for the area	Yes, maps or topographic information, but not detailed	No maps or knowledge of topography
2. Have the hydrologic boundaries of the watershed been delineated and mapped?	Yes, GIS maps of the watershed boundaries and stream reaches in the watershed; delineated zones of protection	Yes, paper maps available	No
3. Have the boundaries of the SWP area (zone(s) of protection) been delineated and mapped?	Yes, and are based on the full watershed or time of travel	Yes, based on arbitrary radial distance from intake (e.g., 5 miles)	No
4. Are data available on time of travel/residence time/flow and precipitation?	Yes, dye study information available and mapped residence time well known for many conditions; flow and precipitation monitored frequently; data easily available	Yes, historical experience on time of travel or residence time from spills; limited flow information, but not frequently used or easily available	No time of travel, time/ residence, time/flow information
5. Is a time-of-travel model available?	Yes, and uses real-time flow data	Yes, based on specific flow values (e.g., 90% of maximum flow)	No
6. Does the water supply depend on a reliable source of snowpack for summer flows?	No	Yes, and the snowpack has generally been consistent over the years	Yes, and the snowpack has generally been decreasing over the years
Groundwater supply			
7. Are maps of the aquifer area available?	Yes, detailed aquifer maps	Yes, simple general maps of region	No
8. Are maps of the geology and descriptions of the aquifer material in the water supply area available?	Yes, electronic/GIS maps of geology that are detailed for the area	Yes, maps or geology information, but not detailed	No maps or knowledge of geology
9. Are the wellhead boundaries defined and mapped?	Yes, maps based on hydrologic information and modeling/time of travel	Yes, paper maps based on fixed radius approach (bulls-eye)	No
10. Are data available on time of travel/monitoring well/ transmissivity?	Yes, detailed estimations based on site-specific monitoring information/flow models; active monitoring well network	Yes, rough estimations based on regional conditions with limited monitoring well information	Not available
11. Is an inventory of the presence of abandoned or improperly closed wells in the SWP area available?	Yes, and relatively confident in the accuracy and completeness	Yes, but only partial or not confident in the accuracy and completeness	None
12. Are the wells and pump station secure?	Yes	Yes, partially	No
13. Is there a reasonable potential for flooding of the wellhead?	No	Possible but not very likely	Yes, has already flooded, or likely possibility of flooding

Worksheet B-2: Water Quality Information

Rationale/Goal

Use this worksheet to determine the status of water quality information and its adequacy for source water characterization. This may point to gaps where data collection efforts need to be bolstered.

Assessment

	High	Medium	Low
1. Are monitoring data of the source water quality at the intake available?	Yes, operational parameters plus specific chemicals or microbials of concern	Yes, basic operational parameters (e.g., pH, alkalinity, turbidity, temp)	No monitoring data available
2. What is the extent of the water quality data at the intake?	Weekly, monthly, quarterly, annual, and long-term data	Sporadic, no extensive compilation of data	No short-term data, no long-term data
3. Are data on event monitoring available (e.g., storm events, recreation impacts?	Yes, multiple sites/occasions	Yes, but for few occasions	None
4. Has information been gathered on historic/known water quality impacts on treatment?	Yes, surveys conducted with treatment staff and other information has been compiled	Yes, but only limited anecdotal information; no concerted effort to compile	No effort made to connect treatment impacts with water quality
5. Has water quality information that is relevant to the water supply area been compiled from other locations/organizations?	Yes, regular data collection or established networks for gathering monitoring data in the watershed (inclusion of Clean Water Act monitoring/total maximum daily loads water impairment data that are available)	Yes, but not on a coordinated or regular basis (sporadic reporting from specific issues or events)	No other data but at the intake
6. Is there a coordinated water quality database for the watershed?	Yes, fully coordinated database with extensive water quality data	Yes, partially coordinated database with limited data collected	No coordinated water quality database
7. Has water quality monitoring been used to identify key areas, subwatersheds, and/or contaminant sources to help target specific goals and activities?	Yes, for all key areas	Yes, for some key areas	Not done

Worksheet B-3: Contaminant Source Inventory Data

Rationale/Goal

Use this worksheet to determine the state of your contaminant source inventory information and its adequacy for source water characterization. This may point to gaps where you need to bolster your knowledge of potential contaminant sources in your water supply area.

Assessment

	High	Medium	Low
1. Linkage of sources to contamination events: Are inventories on record, or is there knowledge of contaminant sources related to contaminant events in the past?	Yes, in-depth and extensive information and is in an easily accessible format	Yes, general information and is in an accessible format	Limited information and is not in an easily accessible format
2. Nonpoint sources: Have land uses been identified in the watershed area?	Yes, detailed GIS land-use coverage with numerous land-use designations (e.g., 30% impervious linked to roads)	Yes, have maps and general designation (rural, urban)	General land-use information on nonpoint sources, but no mapping or designation of specific areas
3. Point source inventories: Is there an inventory of the point sources in the water supply area that could influence the water supply?	Yes, electronic database and locational data on all permitted dischargers, unpermitted dischargers, and permitted industrial points	Yes, major permitted facilities in the water supply area, but incomplete	General information on permitted facilities
4. Have visual watershed surveys been conducted (e.g., windshield surveys)?	Yes, extensive visual survey of practically the whole SWP area	Yes, but limited surveys of relatively small areas	No surveys
5. Has monitoring for major contaminants associated with sources identified in the source inventory been conducted?	Yes, full pollutant scans have been conducted	Yes, analyses for some key contaminants have been conducted	No monitoring conducted

Worksheet B-4: Land-Use Analysis

Rationale/Goal

Use this worksheet to perform a land-use analysis, which is an evaluation of actual or proposed use patterns in the delineated SWP area. Some land uses may increase the risk of potential contamination and/or degradation of source water quality. An evaluation of baseline, current, and future uses of the land can potentially be correlated to changes in water quality over time. Involvement in land-use and zoning decisions provides an opportunity for protecting sensitive areas, limiting incompatible uses, managing development, and similar activities. A local master plan with a 20-year build-out analysis can provide a basis for determining population growth and related stressors and potential future sources of contamination.

Assessment

	High	Medium	Low	None
1. Does the utility control/own land in its SWP area?	Yes, the whole watershed or SWP area	Partial ownership, including sensitive areas such as wellhead areas, reservoir buffers, stream corridors	Partial ownership, but some key sensitive areas are still under private ownership	Little or no ownership
2. Are utility personnel involved in land-use and zoning procedures, ordinances, or decisions?	Yes, well involved and have had a positive impact	Yes, somewhat involved and partially effective	Minimally involved	Not at all
3. Is a historical database of land uses available and accessible?	Yes, records cover the complete history for each source; and electronic versions are available	Yes, archived paper records are available	Partial database	Not available
4. Have future land-use pattern impacts on the source been considered?	Yes, a formal analysis has been completed, including a complete build-out model if appropriate	A partial analysis has been completed	Future land use has been considered, but a formal analysis has not been completed	Not at all
5. Have any land-use protection efforts been implemented (e.g., protected forest land, conservation easements, development covenants, formalized use restrictions, individual agreements with landowners)?	Yes, for the majority to all of the SWP area	Yes, for part of the SWP area	Minimal coverage for the SWP area, but planning is underway	None to date

Worksheet B-5: Physical Barrier Effectiveness Determination

Rationale/Goal

Use this worksheet to determine physical barrier effectiveness (PBE), which is an estimate of the ability of natural geologic materials, hydraulic conditions, and construction features of the well or intake to prevent the movement of contaminants into the drinking water source.

Assessment

For surface water supplies, the PBE considers the size of and retention time in the reservoir, stream, topography, geology, soils, vegetation, precipitation, and groundwater recharge. Refer to question 2 if assessing a surface water source, and question 3 for groundwater sources.

1. Is the PBE determined for the supply?	**High**	**Medium**	**Low**
	Completed	In progress	Not done

2. For surface water, are the following source characteristics identified and catalogued?		
a. Area of tributary watershed	Yes	No
b. Area of water body within watershed	Yes	No
c. Volume of water body	Yes	No
d. Retention time of reservoir (volume divided by average flow)	Yes	No
e. Maximum rate of withdrawal through intake	Yes	No
f. Approximate travel time to the intake for water at the farthest reaches of the impounded water body	Yes	No
g. General topography of the watershed	Yes	No
h. General geology of the watershed	Yes	No
i. General soil type	Yes	No
j. Vegetation cover	Yes	No
k. Mean seasonal precipitation	Yes	No
l. Whether there is significant groundwater recharge to the water body	Yes	No
m. Whether the lake/reservoir stratifies	Yes	No
n. Whether there are water quality problems associated with anoxic conditions	Yes	No

Overall scoring for surface water:	**High** Yes to ≥9 entries	**Medium** Yes to 4–8 entries	**Low** Yes to ≤3 entries

3. For groundwater, are the following source characteristics identified and catalogued?		
a. Wellhead protection area	Yes	No
b. Aquifer recharge area	Yes	No
c. Geology of the aquifer (aquifer material)	Yes	No
d. Hydraulic conductivity or transmissivity of the aquifer	Yes	No
e. Storativity, specific storage, and/or specific yield of the aquifer	Yes	No
f. Whether there is significant recharge to the aquifer	Yes	No
g. Nature of confining layers of the aquifer	Yes	No
h. Overall susceptibility of the aquifer to contamination	Yes	No
i. Do the wells meet well construction standards?	Yes	No

Overall scoring for groundwater:	**High** Yes to ≥6 entries	**Medium** Yes to 4–5 entries	**Low** Yes to ≤3 entries

Worksheet B-6: Intake Structure

Rationale/Goal

Use this worksheet to determine the adequacy of the intake structure operation to provide flexibility to select the best water quality and to protect the water quality before it enters the plant.

Assessment

Is the following information available for the intakes?		
1. What type of intake is present (list the type)?	Yes	No
2. Are multiple intake ports available?	Yes	No
3. Are all intake ports operable?	Yes	No
4. Are intake port selections based on water quality analyses?	Yes	No
5. Are the water quality analyses conducted frequently enough to provide the necessary information for adjusting the selected intakes?	Yes	No
6. Is the area around the intake structure closed to recreational areas?	Yes	No
7. Is aeration used?	Yes	No
8. Are there streams that discharge near the intake or other potential sources of contamination close to the intake that can contaminate the supply?	Yes	No

Overall Scoring: High = Yes to 4 or more entries; Medium = Yes to 2 to 3 entries; Low = Yes to 1 or less	High	Medium	Low

Worksheet B-7: Filling Information Gaps and Needs

Definition/Rationale

As part of the source characterization, use this worksheet to identify information gaps and needs and the steps needed to fill those gaps. Additional monitoring, data collection, research, and literature searches are all methods that can be used to fill these data gaps. These activities are part of the review and revision process for your SWP Program.

Assessment

	High	Medium	Low
1. Have you identified key gaps in information for the source characterization?	Yes, key information gaps were identified, or a thorough characterization was conducted and no key information gaps were found	Some attempt was made to identify information gaps, but not a complete analysis	No attempt made
2. Have you developed methods to collect additional water quality data in areas where gaps have been identified?	Yes, new methods in place to expand water quality data	Discussion and recognition of new methods, but none started	No attempt made
3. Have you developed needed fate and transport information been identified (include research, modeling, literature reviews)? This relates to both basic time-of-travel calculations and to more complex analyses of the potential for contaminants to reach an intake.	Yes, new methods to determine fate and transport information have been identified and initiated	Information on fate and transport has been obtained, but is relatively out of date	No attempt made
4. Have you investigated possible areas to link potential sources of contamination (PSCs) and water quality contamination?	Yes, extensive investigation has occurred to correlate PSCs with identified water quality problems	Minimal investigation has occurred	No investigation conducted
5. Have you tracked relevant legislation and regulations that may have direct or indirect impacts on water quality or the general health of the watershed?	Yes, an active mechanism is in place to track legislation and regulations	Minimal tracking has occurred	No

Worksheet B-8: Analysis of Vulnerability/Susceptibility

Rationale/Goal

Use this worksheet when measuring how to link the watershed areas with water quality and contaminant source information in order to assess the vulnerability/susceptibility of the source water. These exercises help to identify data gaps and prioritize contaminant sources for developing goals for SWP.

Assessment

Assessment	High	Medium	Low
1. Are temporal trends established in water quality parameters to focus on issues of concern?	Yes, temporal trends established, future trends may be projected	Limited temporal trends; limited data available to analyze	No analysis conducted or no data available to conduct analysis
2. Have water quality data from various locations been compared to identify spatial areas of concern?	Yes, spatial areas of concern can be identified	Limited spatial comparisons; limited data available to analyze	No analysis conducted or no data available to conduct analysis
3. Have water quality data been linked with land use/contaminant source inventory data to focus on specific activities that may cause water quality problems?	Yes, quantitative linkages established; water quality modeled	Some qualitative relationships	No relationships established
4. Has the risk of potential contaminating activities in the zones of protection, their distance to the source of supply, and the type and amount of contaminants they store on the premises been considered?	Yes, all three components have been considered	Partial assessment completed	No
5. Have the potential sources of contamination been prioritized in terms of their potential for adversely impacting water quality or for the potential for success in mitigating their potential impact?	Yes	Partial prioritization	No

Worksheet C: Source Water Protection Goals

Rationale/Goal

The development of strategic goals that connect back to the vision and SWP area characterization is an essential step in the creating a successful SWP program. Use this worksheet to prioritize goals to reflect the concerns of greatest importance and also the likelihood of success for different measures, and ideally have qualitative, quantitative, and temporal dimensions.

Assessment

	High	Medium	Low	None
1. Program Goals				
a. Do you have written goals for the SWP program?	Yes	Exist in draft form, but need refinement	Minimal goals have been written	No
b. Are the goals realistically relevant to the overall objective of protecting or maintaining source water quality?	Yes, all goals are directly relevant	Yes, many goals are relevant	A few goals are relevant	Not at all
c. Are the goals prioritized?	Yes, all goals are prioritized	Yes, many goals are prioritized	A few goals are prioritized	Not at all
d. Are the goals SMART (*s*pecific, *m*easurable, *a*ttainable, *r*elevant, and *t*imely)?	Yes, all five aspects have been realized in the goals	Partially so	Minimally so	Not at all
e. Have specific qualitative and/or quantitative metrics for success been identified for each goal?	Yes, specific metrics set for all important goals	Yes, metrics for some of the most important goals set	Just a few goals have assigned metrics	Not at all
f. Has a specific timetable been developed to meet the goals?	Yes, specific times set for all important goals	Yes, times for some of the most important goals set	Just a few goals have a timetable	Not at all
2. Do these goals directly and adequately address the primary existing and future threats to source water quality that were identified in the source water/ SWP area characterization and susceptibility analysis?	Yes, complete match	Yes, partial match	Yes, with minimal coverage	Not at all
3. Do the goals address emerging/unknown contaminants?	Yes, explicitly	Yes, partially	Yes, with minimal coverage	Not at all
4. Do the goals address potential changes in land use and related impacts?	Yes, goals anticipate future land-use change and related impacts (this may include no expected change in land use)	Yes, goals identify some land-use changes but do not directly address related impacts	Goals minimally identify land-use change	Not at all
5. Do the goals address potential impacts from future climate change?	Yes, goals explicitly address potential future issues from climate change	Yes, goals partially consider potential future issues from climate change	Goals barely address potential future issues from climate change	No anticipation of any future climate change issues
6. Do the goals address other potential future issues for the source water?	Yes, goals explicitly address potential future source water issues	Yes, goals partially consider potential future source water issues in the future	Goals barely address anticipated changes to the source water in the future	No anticipation of any future source water issues
7. Do the goals meet or exceed existing regulations?	Yes, goals completely address regulations	Yes, goals address some regulations	Only minimally address regulations	Do not address any regulations
8. Stakeholder involvement				
a. Are internal stakeholders involved in development of the goals?	Yes, extensively	Yes, moderately	Yes, minimally	Not at all
b. Are external stakeholders involved in development of the goals?	Yes, extensively	Yes, moderately	Yes, minimally	Not at all
c. Do these goals adequately consider customer and other stakeholder expectations?	Customer and stakeholder expectations explicitly considered in goals	Customer or stakeholder expectations generally considered in goals	Customer and stakeholder expectations marginally considered in goals	No consideration at all
9. Is there a process for periodic revision and improvement of the goals?	Yes	In development		No

Worksheet D: Source Water Protection Action Plans

Rationale/Goal

Use this worksheet to identify the main basic components of the planning process. Additional detailed evaluation is provided in Worksheets D-1 and D-2.

Assessment

	High	Medium	Low	None
1. Goals and Vision				
a. Does the action plan address the SWP vision?	Yes, all aspects	Yes, some aspects		No
b. Is each of the established SWP goals supported by potential projects and/or activities?	All are covered	Many are covered	Some are covered	None are covered
2. Essential Components				
a. Address existing priority contaminant sources	All sources identified in characterization	Key/most risky sources identified in characterization	In progress	No
b. Address sensitive areas	Yes, all	Yes, some	No	
c. Consider specific actions (e.g., best management practices [BMPs]) for key contaminants, and the likely effectiveness of those actions	Yes, sufficient actions planned and their effectiveness estimated	A few actions planned and their effectiveness estimated	A few actions planned but their effectiveness was not estimated	None planned
d. Involve stakeholders	Yes, stakeholders involved in planning and public comments incorporated	Yes, stakeholders involved in development of key actions	Yes, public comments solicited	No
e. Have the existing SWP programs and initiatives of the key stakeholders and other organizations been identified (including water quality monitoring, BMPs, land conservation, education)?	Extensive identification and compilation of information	Partial identification and compilation of information	Minimal identification and compilation of information	None

Worksheet D-1: Prioritization and Planning

Rationale/Goal

Use this worksheet to identify the primary components of the planning process, including prioritization of activities and development of measures of success.

Assessment

	High	Medium	Low	None
1. Are potential projects and/or activities prioritized on the basis of: • Relative risk from pollutant sources • Buy-in from stakeholders • Staff and resource commitment needed • Budget and finances • Expertise • Time commitments needed to accomplish • Political support and feasibility • Likely effectiveness • Short-term versus long-term actions	All topics are prioritized	Many topics are prioritized	Some topics are prioritized	No topics are prioritized
2. Have work plans been developed for the projects (including scope, budget, required resources, responsibilities, and implementation schedule)?	Yes, all work plans have been developed	Most work plans have been developed	A few work plans have been developed	No work plans have been developed
3. Is a timetable laid out for implementation of each step of the action plan?	Yes, a complete timetable is laid out	A timetable is laid out for most steps	A timetable is laid out for a few steps	No timetable has been laid out
4. Have means for measuring the success of the projects (e.g., identified metrics for monitoring program effectiveness) been developed?	Yes, means for measuring success have been developed for all projects	Means for measuring success have been developed for most projects	Means for measuring success have been developed for some projects	No means for measuring success have been developed
5. Are funding mechanisms in place to support the various potential projects and/or activities?	Yes, funding in place for all high-priority projects/activities	Some funding is in place for some high-priority projects/activities	Minimal funding in place for a small number of the high-priority projects/activities	No funding is in place
6. Have potential problems and obstacles been identified to the extent feasible?	Yes, a comprehensive list of potential problems and obstacles has been identified	Most potential problems and obstacles have been identified	A few potential problems and obstacles have been identified	No potential problems and obstacles have been identified
7. Are there any research efforts to address current and future contamination threats to your source water (including past contamination events)?	Yes, research efforts are underway	Research efforts are under development		No research efforts have been made

	High	Medium	Low	None
8. Does the action plan contain sufficient flexibility to address future needs that may involve: • Pathogenic microorganisms • Nutrients • Taste and odor • Emergency response to hazardous material spills • Water quantity emergencies • Long-term water quantity • Operational/treatment issues • Newly recognized contaminants (e.g., pharmaceuticals and personal care products)	Yes, the plan demonstrates flexibility by anticipating future needs of plan and changes of source water	The plan anticipates some needs but does not consider all possible changes		The plan is insufficient in anticipating future needs
9. Does the plan consider future changes in land use and their impacts on water quality? Was a model used to predict future development impacts?	Yes, and the model is current and updated	Issues discussed and plans laid for future assessment and modeling		Not yet considered
10. Does the plan address consideration of potential future point sources, including how they would be identified?	Yes		No	
11. Does the plan address future sources of supply and how they will be protected? (applies to groundwater supplies, groundwater under the direct influence of surface water, and new intakes on the same watershed)	Yes	Potential future sources identified but protection measures still to be considered	Some future sources considered but concrete plans yet to be made	No future sources considered, and no concrete plans made
12. Is there a process for periodic revision and improvement of the action plan?	A process has been developed	A process for review and improvement is in development		No process has been developed

Worksheet D-2: Contingency Planning

Rationale/Goal

Use this worksheet when developing contingency plans for potential emergency situations, including plans for remediating contaminant spills, assessing security risks, and functioning with the water supply unavailable.

Assessment

	High	Medium	Low	None
1. Has the ability of the water system to function with the loss of the largest source of supply been assessed? • Water system's maximum capacity identified • Capacity re-evaluated to consider if the largest supply source were to be lost • The most vulnerable sources of supply identified (using vulnerability/susceptibility analysis)	Yes, all items in list assessed	Some items in list assessed, others still under consideration	Still to be considered	
2. Has a plan for alternate water supply been developed? • Short-term supplies identified • Long-term supplies identified • Emergency supplies considered, including increasing production from existing supplies, conservation measures, interconnections with other water supply systems, providing standby treatment facilities, increasing storage • Alternative supplies for fire flows considered	Yes, all items in list developed and addressed	Some items in list are still under consideration	Still to be considered	
3. Has a spill/incident response plan been developed? • Included emergency responders (e.g., fire department, police, health agency) in the plan • Included protocols and standard operating procedures (SOPs) dealing with the potential contamination/incident • The necessary equipment (e.g., spill booms) has been obtained and secured • SOPs for sharing information with the media/public	Plan completed and table top and simulated incidents have been conducted	Plan completed but as yet to be tested	Under development	Not yet started
4. Has the utility adequately identified the key security threats to the source water?	Extensive identification	Partial identification	Minimal identification	No identification
5. Does the utility have documentation that describes emergency response plans and provides specific directions to personnel in the event of an emergency (including sabotage and accident)?	Very clear and comprehensive emergency plans, known by all relevant employees	Largely adequate emergency plans, known by all relevant employees	Partially adequate emergency plans, but not known by or available to all relevant employees	Inadequate or no emergency plans
6. Does the emergency plan include components for both protecting people and for protecting the source water?	Yes, for both	Only for one	Neither completed	No effort made
7. Does the utility have documentation of health and safety procedures that are designed to safeguard the employees and visitors engaged in operations activities pertaining to watershed management?	Very clear and comprehensive health and safety procedures, known by all relevant employees	Largely adequate health and safety procedures, known by all relevant employees	Partially adequate health and safety procedures, but not known by or available to all relevant employees	Inadequate or no health and safety procedures

Worksheet E: Program Implementation

Rationale/Goal

Use this worksheet for the implementation of SWP activities to accomplish program goals. A more detailed evaluation is provided in Worksheets E-1 through E-6.

Assessment

	High	Medium	Low	None
Milestones and Achievements				
1. Are the high-priority projects completed or in process?	Yes, all	Yes, most	Yes, some	No
2. Have project milestones been achieved on time?	Yes, all	Yes, most	Yes, some	No
3. Are projects achieving their objectives as outlined in the action plan?	Yes, all	Yes, most	Yes, some	No projects achieved plan requirements
4. Were all components of the plan implemented?	Yes, all components were implemented	Most components were implemented	Some components were implemented	No components were implemented
Roles and Responsibilities, and Responding to Changes and Obstacles				
5. Were there changes of responsibilities or roles of utility personnel during implementation?	No; or Yes and implementation was maintained or improved		Yes, and process was slowed	Yes, and process halted
6. Was there continued support or participation throughout plan implementation by stakeholder partners?	Yes, throughout the whole implementation process and by all partners	Yes, during most of the implementation and by most partners	Yes, during some of the implementation and by some partners	No support from partners during any part of implementation
7. If obstacles to successful implementation of the action plan were encountered, have means for surmounting those obstacles or other means of reaching the objectives been identified?	Yes, for all obstacles	Yes, for most obstacles	Yes, for some obstacles	No
8. Were there any funding changes during implementation of the project?	Yes, funding changes supplemented implementation	No, projects proceeded as planned	Yes, and changes impeded implementation	Yes, and changes halted implementation
9. Is there a process for periodic revision and improvement of the program implementation tasks?	Yes, a comprehensive process is in place	A process is in development		No, there is no process in place or under development

Worksheet E-1: Assessing Key Milestones

Rationale/Goal

An effective SWP program depends on the successful implementation of necessary programs that both prevent and remediate pollution. Use this worksheet when performing a detailed evaluation of the effectiveness of watershed controls, which is essential in assessing whether implementation goals and milestones are being achieved.

Assessment

	High	Medium	Low	None
1. Are controls implemented to adequately address the contaminants of concern previously identified and prioritized in order to minimize degradation and/or contamination of the source water and meet the overall SWP goals?	Yes			No
2. If so, what types of controls and practices are employed? • Regulations/ordinances • Zoning control • Land acquisition • Land management • Land conservation (e.g., easements) • Other land-use agreements • Pollution prevention planning (types) • Agricultural best management practices (BMPs) • Forestry BMPs • Stormwater BMPs • Other BMPs • Development review • Chemical storage containment systems • Local/regional planning • Household and commercial hazardous waste collection • Agricultural chemical collection • Public education • Others	A sustainable mix of controls from the list are employed	A sustainable mix of controls from the list are planned, but only some have been employed	A limited number of controls have been considered and only a very few implemented	No or minimal types of controls have been implemented to date
3. Is there a matrix that shows how the controls address specific contaminants of concern?	Yes, and matrix is complete	Matrix is currently being constructed	This task has not been started	
4. Is progress of the implementation of these controls tracked? If so, how?	Yes, using objective measures in conjunction with a flexible interactive feedback process	Yes, using some objective measures, some anecdotal/ observational measures, with some flexibility for re-evaluation	Using mostly informal and observational measures	No, progress is not being tracked
5. Is SWP served by structural BMPs (e.g., riparian buffers, stormwater basins), and nonstructural BMPs (e.g., land acquisition, zoning restrictions)?	A sustainable mix of structural and nonstructural BMPs are employed	A sustainable mix of structural and nonstructural BMPs are planned, but only some have been employed	Only limited BMP projects have been completed	No BMP projects planned or completed

Worksheet E-2: Roles and Responsibilities

Rationale/Goal

The protection of water sources may involve multiple organizations, government agencies, individuals, businesses, and industries. An effective SWP program may need the ability from both regulatory and statutory standpoints to establish controls as well as solicit backing of the community to accomplish SWP task implementation. Effective programs depend on a well-trained SWP implementation staff and a well-educated public. Therefore, education, training, and the promotion of watershed issues and programs need to be in place for a program to succeed. The lack of these components will limit the effectiveness of the program. Use this worksheet to identify roles and responsibilities.

Assessment

	High	Medium	Low	None
Regulatory Issues				
1. Have all the regulatory concerns surrounding the selected approach been identified and addressed?	Yes, regulatory issues influencing the SWP approach and capacity to deal with them in an effective manner have been identified and addressed	Some issues may require legislation, regulation, or cooperation with outside agencies; however, steps have been taken to address the tasks	Some regulatory obstacles exist and resulted in limitations to success, but others were successfully addressed	Because of regulatory limitations, some aspects of the program have become advisory only and cannot be accomplished directly
Utility Personnel (SWP implementation staff)				
2. Have training needs been identified that will affect implementation success?	Yes, consistent and adequate training provided to a significant number of those in need of such training	Conducted training in a few areas of need	Identified training but have yet to conduct any	No training provided
3. Does training include an evaluation of the training itself?	Yes, and the evaluations are used in subsequent training	Yes	No formal evaluation mechanism is in place, but informal discussions with trainees are used	No evaluation of training
Stakeholders and the Public				
4. Has public participation been included in resolving problems, securing public support for the programs, and assisting in implementation?	Yes, public participation has been included in the problem resolution, and support for the programs is established	Community outreach is being addressed, and public input is anticipated to support the programs	Some, but minimal, effort	Public support has not been sought
5. Is an education program in place in the identified source water area that will contribute to implementation success?	Yes, a comprehensive program addressing all audiences is in place	Program is reaching a subset of audience	Program is passive and available upon request	No program
6. Is there a program to increase awareness of SWP issues and programs in the protection area?	Yes, a strong program	Yes, but not a very active program	No, but planning is underway	No program
7. Are informational brochures or other publications distributed?	Yes, several	Yes, but not many	No, but development and printing has been planned	No
8. Does local media provide ongoing coverage of SWP efforts?	Yes	Sometimes	No, but a plan for such efforts is underway	No
9/ What public participation elements are included in the program? Examples can include: • Local/regional planning • Storm drain stenciling • Children's education • Community festival • Adopt a stream program • Watershed clean-up days • School project competitions (e.g., best poster) • School field trips or monitoring programs • Others	Multiple public participation elements have been planned and implemented	Multiple public participation elements have been planned but have yet to be implemented	Limited public participation elements have been discussed	No public participation elements

Worksheet E-3: Resources

Rationale/Goal

Securing adequate funding for the program is critical to its success, as is having a specific person responsible for the project's progress (a project "champion"). Adequate appropriations, with a sound and secure funding source, must be on hand. Responsibility for the project must be established. The ability to continue to productively implement SWP activities with changing resource availability, both personnel and financial, is critical. Use this worksheet to identify resources.

Assessment

	High	Medium	Low
Financial			
1. Have cost estimates for each key part of the program been established?	Yes, cost estimates have been established	Some costs have been estimated, but more budgeting work is needed	Cost estimates have not yet been developed
2. Has adequate funding been secured?	Yes, adequate funding has been acquired	Funding sources have been identified but not yet secured	No funding sources yet identified
3. Have alternative possible sources of funds been identified should the financial situation change?	Yes, alternative possible sources of funds have been identified		No, alternative possible sources of funds have not been identified
Personnel			
4. Has an individual within the organization been identified as the responsible party for the overall SWP program?	Yes, an assignment, with authority to complete the project, has been given to a responsible individual	No specific assignment has been made, but there is backing for the project within the organization, with the intention to allocate staff	No specific individual is responsible for the program
5. Are enough qualified personnel available and dedicated to the SWP effort?	Yes, there are enough qualified personnel available and dedicated to the effort	Some staff assigned, but adequacy uncertain	No, there are no qualified personnel available and dedicated to the effort
6. Have cross-training and succession planning for personnel changes been established?	Yes, cross-training and succession planning established	No specific plans, but options have been explored	No cross-training and succession planning

Worksheet E-4: Water Quality Monitoring

Rationale/Goal

    Water quality monitoring should include adequate breadth of contaminants and frequency of analysis to provide critical information about the status of source water quality. Periodic review of monitoring data over time can provide insights into gradual changes in water quality that may be watershed dependent. Use this worksheet to evaluate water quality monitoring.

Assessment

	High	Medium	Low
1. Is a water quality monitoring plan/schedule in place to assess water quality and evaluate changes in the watershed?	Yes, and it includes a suite of potential contaminants based on vulnerability assessments in addition to routine monitoring	Yes, primarily routine monitoring	No monitoring plan/schedule in place
2. Is sampling conducted during/after precipitation events to capture runoff events?	Yes, a formalized process exists	Yes, on an occasional basis	No sampling conducted
3. Are water quality data compared from year to year to track changes?	Yes, a formalized process and schedule exists	Yes, on an informal basis	No, water quality data are not compared

4. Is special monitoring conducted to track impacts of activities in the watershed?
  • Hydrologically based sampling with an upstream/downstream design
  • Application of microbial source tracking tests
	Yes, for a majority of sensitive areas or sources of concern	Yes, on an incident-triggered basis	No special monitoring conducted

5. Are the data used in an attempt to evaluate the impacts of best management practices and other protection practices?	Yes, for a majority of sensitive areas or sources of concern	Yes, on an incident triggered basis	No data for evaluation

Worksheet E-5: Biological/Habitat Monitoring

Rationale/Goal

Species composition and diversity can serve as a real-time sentinel of watershed system health. Habitat diversity can provide mitigation of anthropogenic impacts on water quality. Ongoing evaluation of biological and habitat health of watersheds can also be useful in evaluating the effectiveness of your watershed plan. Use this worksheet for biological/habitat monitoring.

Assessment

	High	Medium	Low
1. Do you have a watershed biological/habitat monitoring program, or do you work with other agencies to monitor biological/habitat conditions?	Yes, and it is relatively thorough	Yes, but limited	No program
2. Have you established criteria to evaluate the effectiveness of your watershed biological/habitat monitoring program?	Yes, quantitative measures and qualitative benchmarks are used	Yes, informal measures are used	No criteria
3. How is effectiveness measured? • Water quality improvement • Habitat improvement • Aquatic species diversity measures • Terrestrial species diversity measures • Number of best management practices (BMPs) installed/maintained/inspected • Percent of area protected by BMPs	By a combination of three or more components listed	Using at least one of the components listed	No formal measures
4. If watershed controls for invasive species are used (i.e., hemlock wooly adelgid, emerald ash borer, mountain pine beetle), is there a mechanism in place for evaluating program effectiveness?	Yes, quantitative measures and qualitative benchmarks are used	Yes, informal measures are used	No mechanisms in place
5. Have biological indices been adopted for measuring anthropogenic influences (such as land-use changes, contamination events, and potential sources of contamination) and evaluating watershed health?	Yes, biological indices are used to assess influences and evaluate watershed health	Using mostly informal and observational measures	No indices adopted

Worksheet E-6: Stakeholder and Public Relations

Rationale/Goal

Involvement of various stakeholders and the public should be on-going and renewed in order to bolster buy-in and support for watershed protection activities. Use this worksheet to evaluate stakeholder and public relations.

Assessment

	High	Medium	Low	None
1. Does a process exist for involving stakeholders in your SWP program evaluation/ updating process?	Yes, extensive efforts are made to incorporate stakeholder input	Yes, a formal process exists	Public comment has been solicited	No process exists
2. Is the SWP plan updating process transparent and equitable?	Yes, based on stakeholder evaluations	Yes, formal processes to involve stakeholders are used	Informal stakeholder involvement is currently used	No process in place
3. Does the evaluation process provide an atmosphere or culture of using trust, teamwork, collaboration, and equity to achieve results?	Yes, based on stakeholder evaluations	Yes, based on utility personnel evaluation	Stakeholder involvement is included, but have no basis for judgment	No stakeholder involvement
4. For activities and educational programs aimed at the general public, is a mechanism in place to solicit feedback from recipients?	Yes, a formal process for feedback is in place	Yes, informal comments on program effectiveness are solicited	No formal or informal feedback mechanisms are in place to date	
5. For land-use restriction agreements, development covenants, and individual agreements with land owners, is there a mechanism to evaluate their effectiveness and support from stakeholders?	Yes, a formal process for feedback is in place	Yes, informal comments on program effectiveness are solicited	No formal or informal feedback mechanisms are in place to date	

Worksheet F: Program Evaluation and Revision

Rationale/Goal

Administrative programs of any type require periodic (or continuous) evaluation and revision. Use this worksheet to evaluate the basic concepts of this practice.

Assessment

	High	Medium	Low	None
Review Process				
1. Is there an established process for periodically evaluating the SWP program, including the vision, source characterization, goals, work plans, and implementation steps?	Yes, formal process is planned and implemented as scheduled	Yes, formal process is planned but not always implemented as scheduled	Process is planned but not yet implemented	No planned process
2. Is there an established timeline and/or other criteria for determining when to conduct evaluations of the program?	Yes, for the entire process	Yes, for much of the process	Yes, for some of the process	No timeline or other criteria
3. Is there an established process for revising the SWP program based on the results of the evaluation?	Yes, for the entire program	Yes, for much of the program	Yes, for some of the program	No process in place
4. Is a process in place to identify and assess emerging issues and changing land-use practices and incorporate them into the program?	Yes, a comprehensive process is in place	Some process is in place	A process is under development	No process in place
Roles and Responsibilities				
5. Has the party responsible for evaluating the SWP program been named?	Yes, the party has been named	Not specifically, but personnel are available		No party has been named
6. Is stakeholder involvement used for the evaluation?	Yes, extensively	Yes, moderately	Yes, but minimally	Not at all
7. Is the SWP program evaluation and modification reported to internal and external stakeholders and the governing board?	Yes, a comprehensive evaluation report is given	A partial evaluation report is given	No	No
Benchmarks				
8. Have benchmarks against which to gauge program progress been established? • Land-use assessment? • Water quality monitoring? • Biological/habitat monitoring? • Land management practices? • Best management practices implementation? • Education programs? • Community and stakeholder communications? • Other areas of the SWP program?	Yes, for all aspects of the program	Yes, for most aspects of the program	Yes, for some aspects of the program	No benchmarks
9. Based on benchmarks, is there a need for additional source water area characterization (potential sources of contamination identification, data, monitoring, measures, stakeholders)?	Yes, and all additional data have been collected	Yes, and most additional data have been collected	Yes, and some additional data have been collected	Not determined
10. Based on benchmarks, is there a need for additional SWP activities?	Yes, and all additional data have been collected	Yes, and most additional data have been collected	Yes, and some additional data have been collected	Not determined
11. How does the performance of the various projects measure against their established targets or goals?	Performance for all projects matches targets/goals	Performance for most projects matches targets/goals	Performance for some projects matches targets/goals	Performance for no projects matches targets/goals

Worksheet G: Verification and Record Keeping

Rationale/Goal

Appropriate management and maintenance of adequate records and documents is essential to the success and efficient progress of any program, as well as to verification of compliance with this Standard G300. Use this worksheet to evaluate verification and record keeping.

Assessment

	High	Medium	Low
1. Is a document management system in place?	Yes, and it is relatively thorough	Yes, but limited	No document management system
2. Does the utility have documented, established, and implemented policies and procedures for adequate control of records and documents, including their identification, storage, protection, retrieval, retention time, and disposal, as appropriate?	Yes, a specific policy and SOP are in place	Yes, informal measures are used	No specific policy or SOP
3. Are records legible, readily identifiable, and retrievable?	Yes, well organized and readily available for staff	Yes, but limited in organization and retrievability	Not well organized or easily retrieved
4. Are the following records maintained and retrievable? • Executed resolutions and recorded minutes of the utility's governing body • Summaries or minutes of relevant advisory committee or stakeholder meetings • Summaries or minutes of relevant public hearings • Technical studies • Monitoring data	All or most of the components listed	Some of the components listed	Few or none of the components listed
5. Is an adequate source water quality data and information management system in place? This would include meta-data (i.e., data about data) such as chain-of-custody and quality assurance/quality control data.	Yes, and it is relatively thorough	Yes, but it is limited	No system in place
6. Is documentation that describes the SWP program goals, the action planned to meet those goals, the actions performed to meet those goals, and any measurements or other methods made to gauge the progress and success of the action items readily available?	Yes, and it is relatively thorough	Yes, but it is limited	No documentation
7. Does the utility have all the critical documents identified in this standard readily available?	Yes, for most or all components	Yes, for some components	Yes, for few or none of the components
8. Does the utility maintain adequate records to provide evidence of conformity and of the effective operation and implementation of the standard?	Yes, and it is relatively thorough	Yes, but it is limited	No records maintained

SECTION 11: CASE STUDIES FOR SOURCE WATER PROTECTION

Case Study: Philadelphia Water, Philadelphia, Pennsylvania

Introduction to Philadelphia Water's Source Water Protection Program

Philadelphia Water serves the greater Philadelphia region, providing integrated water, wastewater, and stormwater services. Philadelphia Water provides an average of 260 million gallons per day (mgd) of potable water to more than 1.7 million people in the city of Philadelphia and surrounding suburban areas. Philadelphia Water's water supply comes from the Schuylkill and Delaware rivers. Schuylkill River water is treated at either the Belmont Water Treatment Plant (WTP) or the Queen Lane WTP, with a combined average production of approximately 105 mgd. Delaware River water is treated at the Baxter WTP, with an average production of more than 155 mgd. All three drinking WTPs are located near the bottom of these two very large and diverse watersheds, as shown in Figure CS1-1.

The Delaware River is 330 miles long with more than 2,000 tributaries. From the forested headwaters in the Catskill Mountains of New York to its mouth, an industrial international seaport at the Delaware Bay, the river drains 13,600 square miles in four states: New York, Pennsylvania, New Jersey, and Delaware. More than 15 million people, including populations in New York City and Philadelphia, rely on the Delaware River watershed for drinking water from ground and surface water sources (DRBSWC 2016).

The Schuylkill River is more than 130 miles long, includes more than 180 tributaries, and drains an area of approximately 2,000 square miles (PWD 2002a). It is the largest tributary to the Delaware River. The river and streams in the Schuylkill River watershed are the drinking water source for more than 2 million people (SAN 2013). The watershed land is mixed use—approximately 37% forested, 28% agricultural, and 27% developed (PWD 2015b).

Source water protection (SWP) is a daunting task in such large watersheds where the drinking water utility does not have ownership of the water resources or the surrounding land. In this case, the key to Philadelphia Water's successful Source Water Protection Program is a watershed approach fueled by partnerships and stakeholder collaboration. Since its inception, the SWP Program has operated

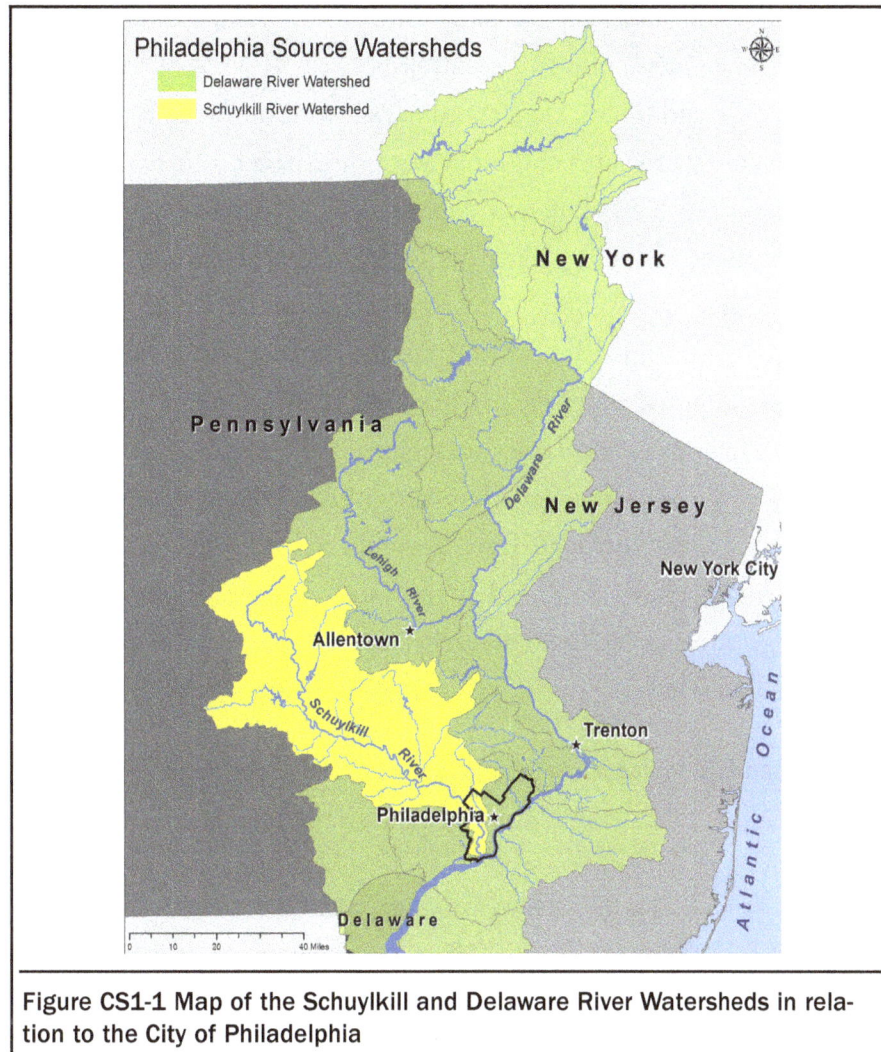

Figure CS1-1 Map of the Schuylkill and Delaware River Watersheds in relation to the City of Philadelphia

in a cooperative environment with an expansive and diverse group of watershed stakeholders who work together to achieve common environmental outcomes. The positive support from Philadelphia Water executives and city officials as well as fellow stakeholders has contributed to Philadelphia Water's successes and has led to increased SWP opportunities.

Established in 1999, Philadelphia Water's SWP Program was inaugurated as a component of a larger initiative at Philadelphia Water to protect Philadelphia's water resources. This was achieved by combining their SWP Program, Stormwater Program, and Combined Sewer Overflow Program under a single Office of Watersheds. Staff from each program coordinate on projects to reduce overlapping protection efforts, collaborate on initiatives, and improve results from programs driven by the Clean Water Act and the Safe Drinking Water Act. Information on the various Philadelphia Water Office of Watersheds programs is available on their website (www.phillywatersheds.org).

In addition to close coordination with the Stormwater and Combined Sewer Overflow programs and other watershed-based initiatives in the Office of Watersheds, the SWP Program coordinates internally with other Philadelphia Water programs including Water Treatment and Operations, the Bureau of Laboratory Services, and Planning and Research. Through collaborative efforts with these programs, Philadelphia Water makes connections between watershed conditions upstream and water quality observed at the drinking WTPs. Philadelphia Water also follows federal, state, and regional policies that affect water quantity and quality in the Schuylkill and Delaware rivers and analyzes water quality data for long-term trends. Anticipated changes and challenges upstream are considered in utility planning efforts for Philadelphia Water.

Some key milestones reached by Philadelphia Water's SWP Program are listed in Table CS1-1.

Table CS1-1. Milestones of Philadelphia Water's Source Water Protection Program

Year	Milestone
1999–2003	• Conducted source water assessments for all intakes in the Schuylkill watershed and most intakes on the lower Delaware River
2002	• Awarded a SWP grant by the Pennsylvania Department of Environmental Protection
2003	• Established the SAN
2003–2006	• Developed SWP plans for the Schuylkill and Delaware rivers
2004	• Awarded a $1.15 million USEPA Targeted Watershed Initiative grant for the Schuylkill River with the PDE
2004–present	• Implemented their SWP plan
2005–present	• Implement, maintain, and continue to develop a regional web-based early warning system
2005–present	• Established and maintained a web-based bacteria forecasting system for recreational users of the Schuylkill River
2006–present	• Conducted extensive research to identify sources of *Cryptosporidium* using state-of-the-art DNA analysis
2007–present	• Supported the Special Protection Waters Program with the Delaware River Basin Commission
2007	• Developed a model to identify the most valuable lands for Philadelphia water supply protection in the Schuylkill watershed
2007	• Actively began participating as technical advisor to the State of Pennsylvania to support ongoing water allocation negotiations among the parties to the 1954 Supreme Court Decree
2008	• Completed more than 35 projects in the Schuylkill watershed through the Targeted Watershed Initiative Grant Program with PDE and other SAN partners
2008–2012	• Completed pharmaceutical take-back pilot program
2008	• Reviewed natural gas extraction in Pennsylvania's Marcellus shale region and provided public comments when possible
2009	• Incorporated surface water–dependent industries into the regional early warning system
2010	• Performed a one-dimensional modeling analysis of the Delaware estuary to simulate the relationship between salinity and freshwater streamflow in the vicinity of the Philadelphia Baxter drinking water intake
2010	• Developed a water budget for the Schuylkill watershed to observe, at a planning level, the interactions between consumptive use, hydrology, and drinking water supply availability

2010	• Became an annual contributor to the Schuylkill River Restoration Fund
2011–2013	• Led a regional iodine-131 watershed characterization study and corresponding education and outreach initiatives for water professionals and customers
2012	• Received State of Pennsylvania approval for a watershed control plan for compliance with the Long-Term 2 Enhanced Surface Water Treatment Rule
2013	• Celebrated the 10-year anniversary of the SAN
2013	• Developed novel drinking water supply availability metrics and completed an operational analysis and simulation of integrated systems (OASIS) model of the Schuylkill River watershed to analyze watershed management policy outcomes on the Philadelphia drinking water supply
2014	• Developed and incorporated a tidal spill trajectory tool into the early warning system using a Port Security grant awarded to Philadelphia Water by the Maritime Exchange for the Delaware River and Bay
2014	• Received recognition in the Philadelphia Water Strategic Plan as a key component of the strategic objective to ensure sustainable utility operations
2015	• Completed a watershed sanitary survey as required for the Watershed Control Plan credit for compliance with the Long-Term 2 Enhanced Surface Water Treatment Rule

Abbreviations: PDE, Partnership for the Delaware Estuary; SAN, Schuylkill Action Network.

Philadelphia Water's SWP Program has received several awards in recognition of the outstanding efforts made for their source water assessment (SWA) and protection work. Those awards include the following:

- USEPA Region III Source Water Protection Award—2002
- USEPA Clean Water Partner for the 21st Century—2003
- AWWA Exemplary Source Water Protection Award—2003
- Schuylkill Watershed Community Partner Award—2013, Delaware Valley Early Warning System
- Pennsylvania's Governor's Award for Environmental Excellence—2013, Schuylkill Action Network (SAN)
- Pennsylvania's Governor's Award for Environmental Excellence—2015, Delaware Valley Early Warning System

Source Water Protection Program Vision and Stakeholder Involvement

Philadelphia Water has enthusiastically embraced the role of regional "watershed champion," accepting the challenge of implementing a comprehensive environmental approach to water resource management. The Philadelphia Water vision is perhaps best exemplified by the formation of the Office of Watersheds in January 1999. The Office of Watersheds integrates three historically separate programs: Combined Sewer Overflow, Storm Water Management, and Source Water Protection. This reorganization was intended to optimize the resources allocated to controlling the City of Philadelphia sewer discharges and protecting drinking water resources to ensure the comprehensive achievement of regulatory requirements within the guiding framework of watershed management.

Figure CS1-2 Conceptual design of Green City, Clean Waters green stormwater infrastructure implementation on a residential street (PWD 2011a)

The Office of Watersheds is nationally recognized for its Green City, Clean Waters initiative, a 25-year plan to protect and improve the health of Philadelphia's rivers and creeks through innovative green infrastructure stormwater management. The vision is "to unite the City of Philadelphia with its water environment, creating a green legacy for future generations while incorporating a balance between ecology, economics and equity," as shown in the conceptual rendering of a green street in Figure CS1-2 (PWD 2011a). The Green City, Clean Waters mission statement is "to preserve and enhance the health of the region's watersheds through effective wastewater and storm water services and the adoption of a comprehensive watershed management approach that achieves a sensible balance between cost and environmental benefit and is based on planning and acting in partnership with other regional stakeholders" (PWD 2009).

AWWA/ANSI G300 Standard requirements for Source Water Protection Program Vision and Stakeholder Involvement are as follows (AWWA 2014):

4.1.1   Vision. The utility shall have a vision or policy that expresses a commitment to source water protection. The vision or policy (or other similar utility document, or consistent practice) shall include a commitment of, or intention to commit, sufficient resources to the source water protection effort. A written version of the vision is strongly encouraged because it would serve to preserve institutional history and allow future generations to be reminded of the vision, along with the opportunities to review and update the vision.

4.1.2   Stakeholder Involvement. Involvement of relevant outside stakeholders is usually essential for development and implementation of a successful source water protection program. The utility shall identify source water area stakeholders, their roles, and existing initiatives in which they may be engaged. Cooperation or partnerships with relevant stakeholders shall be realistically assessed and actively pursued through program development and implementation. The formation of collaboratives at the watershed and local levels is an effective approach to engage stakeholders and partners. Stakeholder involvement may result in improved coordination of partnership activities, additional volunteer efforts, and potential funding opportunities. It is expected that various stakeholders may be involved in each stage of the source water protection program process, but not all of the same stakeholders may be involved in each of the stages.

The Philadelphia Water vision is to exemplify the 21st Century urban water utility, one that fully meets the complex responsibilities and opportunities of our time and environment. A core mission identified in the 2014 Philadelphia Water Strategic Plan is to protect the environment by managing and treating wastewater and stormwater, protecting and advocating for rivers and streams and their watersheds, and protecting sources of drinking water (PWD 2015c). Philadelphia Water continues to provide sufficient resources to implement these goals through their office and watersheds programs and specific SWP Program initiatives.

Characterization of Source Water and Source Water Protection Area

AWWA/ANSI G300 Standard requirements for Characterization of Source Water and Source Water Protection Area include Secs. 4.2.1, Delineation; 4.2.2, Water quality and quantity data; 4.2.3, Contaminant sources, land use, and other threats; and 4.2.4, Inventory of Regulations (AWWA 2014).

Between 1999 and 2003, Philadelphia Water participated in the Pennsylvania Department of Environmental Protection (PADEP) Source Water Assessment Program as the primary contractor for surface water suppliers in the Schuylkill River and Delaware River watersheds. Through this process, Philadelphia Water and its partners developed an appreciation for the many challenges to SWP in such large and diverse watersheds.

Through the Source Water Assessment Program, Philadelphia Water performed an extensive evaluation and characterization of the watersheds for their respective intakes (PWD 2002a, 2002b). Philadelphia Water also completed SWAs for other water utility intakes on the Schuylkill and Delaware rivers and, as a result, collected information about all areas of the watersheds. USEPA Region III recognized the high quality of these SWAs when they presented Philadelphia Water with their 2002 Source Water Protection Award.

The SWA reports for the Schuylkill and Delaware rivers are comprehensive and include detailed analyses on the following areas:

- Watershed characteristics and history
- Key findings from related watershed studies
- Identification of universal water quality issues
- Assessment of current and future water quality monitoring needs
- Inventory and prioritization of potential point sources of contamination
- Identification of previous and ongoing restoration efforts
- Role of public participation in conducting the SWAs, and
- General recommendations for watershed protection.

Recommendations included in the SWA served as a framework to develop the Source Water Protection Plan (SWPP). General recommendations from the SWA pertained to securing grant funding for watershed protection projects, supporting watershed organizations, and protecting and preserving priority SWP areas. Actions identified for focus of SWP efforts included regulatory enforcement of sewage discharge; mitigation of stormwater runoff impacts, including runoff from urban areas, agricultural land, abandoned mine drainage, minimization of erosion, and sedimentation; and wildlife impacts. Recommendations also focused on communication and emergency response during spills and accidents, public education and outreach, data and informational needs, and water quality monitoring for protection and assessment efforts.

A watershed build-out analysis was conducted as part of the Philadelphia Water SWPP, providing additional insight beyond that given by the SWAP report.

The build-out analysis was completed using the US Environmental Protection Agency (USEPA) Source Water Management Model and available county zoning data. Where local zoning data were not available, Philadelphia Water assumed the zoning to be rural low-density residential, which overestimates likely development. Philadelphia Water determined the estimated change in land use and the resulting increase in contaminant runoff that would result from complete build-out in the watershed.

Upon completion of the SWA, the next challenge was to determine how to use the knowledge gained through the assessment process to initiate coordinated and collaborative protection efforts that would address the priorities identified. Numerous potential pollution sources in the watersheds present a challenge for prioritizing SWP activities. For example, the Schuylkill River SWA identified 3,332 point sources/regulated facilities, 76 wastewater plants, hundreds of farms, numerous abandoned and active mines, numerous nonpoint sources, and potentially many other unidentified sources (PWD 2002a). The approach to prioritize and address these challenges is documented in the Schuylkill and Delaware SWPPs (PWD 2006, 2007).

Philadelphia Water's efforts to characterize and better understand its watersheds did not end with completion of the SWPP and continues today. In March 2010, Philadelphia Water completed a water budget for the Schuylkill River watershed in order to better gauge drinking water availability now and into the future. The program has also expanded to anticipate future influences on the water supply system, including the impact of climate change, emerging contaminants, and potential energy development in the basin. In December 2012, PADEP approved the Philadelphia Water Watershed Control Plan (WCP) for compliance with the Long-Term 2 Enhanced Surface Water Treatment Rule (LT2-ESWTR). The WCP identifies potential and actual sources of *Cryptosporidium* in the designated area of influence for the Philadelphia Water Queen Lane drinking water intake on the Schuylkill River. It evaluates the effectiveness and feasibility of various control measures to address priority sources of *Cryptosporidium*, establishes a set of goals for implementation over a five-year period, and presents a quantitative assessment of the planned control measures (PWD 2011b).

AWWA/ANSI G300 Standard requirements (AWWA 2014) for Secs. 4.2.1, Delineation; 4.2.2, Water quality and quantity data; and 4.2.3, Contaminant sources, land use, and other threats are documented in detail in Philadelphia Water SWA reports (PWD 2002a, 2002b) and WCP report (PWD 2011).

Philadelphia Water addresses Sec. 4.2.4, Inventory of Regulations, through its Planning and Environmental Services Division. In 2014, the SWPP sought input from utility stakeholders, including Philadelphia Water treatment plant managers, laboratory managers, and other water utility decision makers. Input from utility stakeholders was then used to compile a comprehensive list of relevant state and federal regulations to evaluate past, present, and anticipated future compliance challenges and ultimately set water system performance metrics. This evaluation was used to develop a regulatory prioritization planning matrix, which will inform system planning efforts.

Source Water Protection Goals

According to AWWA/ANSI G300 Standard requirements, goals should address water quality issues such as public health and aesthetic concerns but also may include other considerations, such as environmental stewardship, biological diversity, socioeconomic and political equity, trade-offs with competing policy objectives (such as transportation, housing, economic development), and others. Goals may address both current and potential future issues. Both internal and external stakeholders should be included in the development of the goals.

AWWA/ANSI G300 Standard requirements for SWP goals are as follows (AWWA 2014):

4.3.1   *Program goals.* The utility shall have written goals for the source water protection program. The utility shall include stakeholders in the development of the goals and shall document that involvement. At a minimum, the goals should
   a. Address the specific problems or issues identified in the source water characterization element;
   b. Be expressed in terms that can be measured or otherwise evaluated in the future; and
   c. Meet or surpass existing and pending regulations, and provide the flexibility to incorporate future regulatory compliance.

The goals outlined in the SWPPs for the Schuylkill River (PWD 2006) and the Delaware River (PWD 2007) meet each AWWA/ANSI G300 Standard requirement, and the plans include substantial detail for each area.

The SWPP for the Schuylkill River incorporates the following seven major objectives (PWD 2006):

- Establish the SAN as a permanent watershed-wide organization charged with identifying problems and prioritizing projects and funding sources to bring about real improvement in water quality throughout the Schuylkill River watershed.
- Create a long-term, sustainable fund to support restoration, protection, and education projects in the Schuylkill River watershed.
- Increase awareness of the Schuylkill River watershed's regional importance as a drinking water source.
- Initiate changes in policies and decision-making that balance and integrate the priorities of both the Safe Drinking Water Act and Clean Water Act.
- Establish the early warning system as a regional information sharing resource and promote its capabilities for monitoring water quality and improving emergency communications.
- Reduce point source impacts to water quality.
- Reduce nonpoint source impacts to water quality.

Numerous metrics for measuring success in meeting each goal are defined in the Schuylkill River SWPP (PWD 2006, pp. 190–199).

While the Philadelphia Water approach in the Schuylkill River watershed has been to create a new organization that works toward common goals (the SAN), the approach in the Delaware basin has been to use established organizations such as the Delaware River Basin Commission (DRBC) and the Delaware Estuary Program to address drinking water concerns. These two different approaches are due in large part to differences in the size of the two watersheds, stakeholder makeup, and priorities.

The goals outlined in the SWPP for the Delaware River (PWD 2007) also meet each AWWA/ANSI G300 Standard requirement. The SWPP for the Delaware River incorporates the following four major objectives (PWD 2007):

- Ensure the Baxter WTP is adequately protected under regional water policy from climate change effects on the salt line and streamflow.
- Prevent the Baxter Water Treatment Plant from losing bin 1 status under the LT2-ESWTR.
- Become a regional leader and facilitator of efforts to offset the effects of land cover change on the water quality and quantity of the Delaware River.
- Raise the profile of the Delaware River as a drinking water supply that needs to be maintained and protected in the eyes of the public, government, and regulatory communities.

Several additional goals for the Delaware River have been added to the program since the SWPP was written in 2007:

- Set specific quality and quantity goals for the water supply.
- Evaluate alternative sources if goals cannot be met.
- Better integrate the Philadelphia Water SWP Program with other Philadelphia Water units (i.e., treatment and conveyance) in order to develop a single comprehensive drinking water program.
- Establish and promote a policy agenda that is focused on protection of headwaters and flow management in order to support achievement of established quality and quantity goals.
- Establish watershed monitoring and data sharing to measure progress against water quality and quantity goals.

Metrics for measuring success in meeting these goals for the Delaware River SWPP have been established as appropriate. For example, to ensure the Baxter WTP is adequately protected under regional water policy from salinity intrusion, Philadelphia Water established a preliminary goal of maintaining chloride at or below 50 mg/L at the Baxter intake. To prevent the Baxter WTP from losing bin 1 status under the LT2-ESWTR, metrics are defined by the <0.75 oocysts/L level for bin 1 as defined by the LT2-ESWTR. To lead and facilitate efforts to offset effects of land cover change, Philadelphia Water is working with watershed stakeholders to define specific land areas in the basin that are most important to water supply protection. Furthermore, several goals outlined in the Delaware SWPP describe the need for further study that will support metrics development.

The WCP aims to address *Cryptosporidium* and pathogens in the Schuylkill River watershed for LT2-ESWTR compliance. The WCP objectives further exemplify AWWA/ANSI G300 Standard requirements for SWP goals. The WCP outlines goals to do the following:

- Adequately address all high-priority sources of *Cryptosporidium* within city limits. Within city jurisdictional boundaries, Philadelphia Water is able to directly implement SWP measures.
- Address high-priority sources of *Cryptosporidium* throughout the entire watershed. Outside of Philadelphia, Philadelphia Water supports SWP through the watershed-wide initiatives of various partnerships.
- Educate city and watershed residents as to various SWP issues, including pathogen contamination.

- Support research initiatives that will further develop Philadelphia Water's understanding of the role of animal vectors in the fate and transport of *Cryptosporidium* throughout the watershed.

Philadelphia Water has developed metrics for measuring their success in meeting these goals, which are detailed in the WCP (PWD 2011b). Additionally, supporting research to better understand *Cryptosporidium* in the watershed surpasses the current requirements and can inform future WCP initiatives for Philadelphia Water.

Action Plan

An action plan, as described in AWWA/ANSI G300 Standard, identifies required actions (e.g., management practices, statutory or regulatory changes, agreements) needed to mitigate existing and future threats to source water quality and establishes priorities and a timetable for the plan's implementation. Specific projects, programs, and/or other activities needed to achieve each SWP goal should be identified. Furthermore, potential projects, programs, and activities should be prioritized, as appropriate, on the basis of their likely effectiveness, availability of necessary resources, timing, stakeholder buy-in, political feasibility, and other considerations.

Part of AWWA/ANSI G300 Standard requirements for the SWP action plan are as follows (AWWA 2014):

Sec. 4.4   Action Plan

The action plan identifies required actions (management practices, statutory or regulatory changes, agreements, and so on) needed to mitigate existing and future threats to source water quality. It establishes priorities and sets a timetable to implement source water protection goals. The action items in the plan shall include the following:

4.4.1   Specific projects, programs, or other activities needed to achieve each of the source water protection goals shall be identified.

4.4.2   Specific projects, programs, and activities shall be prioritized, as appropriate, based on their likely effectiveness, availability of necessary resources, timing, stakeholder buy-in, political feasibility, and other considerations.

4.4.3   Necessary resources shall be identified, such as staff, funding, special expertise (e.g., police, health department and fire department), and cooperation and partnerships with stakeholders, and provisions for obtaining them shall be included in the plan.

4.4.4   Potential barriers or obstacles to the action plan's implementation shall be acknowledged, and provisions for resolving them shall be incorporated into the plan.

4.4.5   Controls to monitor project/program progress, to document progress and successes, and to monitor funding or budgetary changes shall be identified.

The Philadelphia Water SWP Program takes these various factors into account and specifically describes numerous projects that are intended to reduce the impacts of priority contaminant sources. Separate SWPPs were developed for Philadelphia Water's two major watersheds (PWD 2006, 2007). These action plans include prioritized lists of projects and initiatives that specifically address each primary program goal, along with timelines and budgetary and funding considerations for each. Philadelphia Water used computer modeling to support prioritization of potential contaminant sources and related SWP projects.

Restoration and protection projects for the Schuylkill River were identified based on a prioritization analysis that built on the results of the SWAs, build-out analysis, various other river assessment reports, PADEP 303(d) stream assessments, total maximum daily load status, and recommendations of the SAN. The Delaware River SWPP (PWD 2007) also used a detailed methodology to prioritize the various point and nonpoint sources of pollution and sources of specific contaminants. That work supported the development of overall goals as well as prioritized lists of specific projects and initiatives. Several research efforts were identified to assist in better identifying priority pollutant sources and protection activities for various contaminants.

As one response to the need to move from assessment to protection, Philadelphia Water, in cooperation with USEPA Region III and the PADEP, established the SAN to help protect and restore Schuylkill waters through partnerships with organizations, businesses, and governments. The SAN directly addresses the following four major threats identified in the SWA: abandoned mine drainage, agriculture, sewage discharges, and stormwater. For each threat, the SAN created a workgroup charged with prioritizing specific sources and identifying projects for managing them. The SAN is led by the Executive Steering Committee and the Planning Committee. Additional workgroups have also been created, including a Watershed Land Protection Workgroup and an Education and Outreach Workgroup (Figure CS1-3). Information on the SAN can be found at www.schuylkillwaters.org.

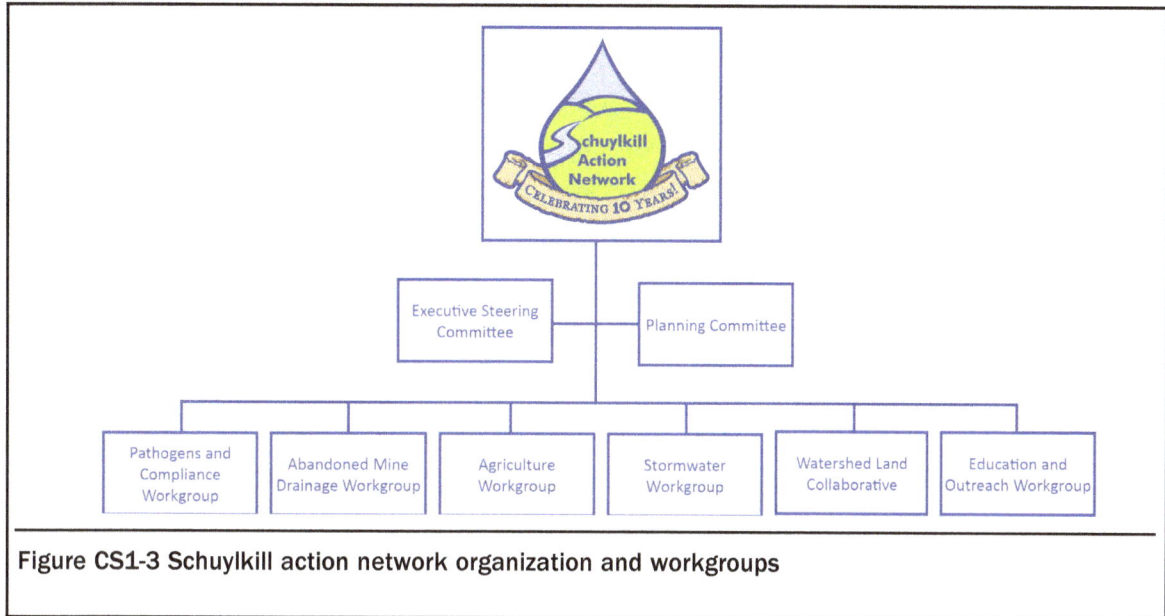

Figure CS1-3 Schuylkill action network organization and workgroups

The SAN has been very effective and serves as a model for other collaborative multistakeholder watershed initiatives. Philadelphia Water and the SAN have prioritized and initiated numerous projects to address significant point and nonpoint sources of contamination. Monitoring programs have been developed to assess the efficacy of as many of these projects as practical. Philadelphia Water's support for this effort has been notable, as they designated and partially support a full-time position to coordinate SAN efforts. Coordination of the various SAN stakeholders' efforts is impressive, and progress has been made on many fronts. The concept that the SAN serves as a "one-stop shop" for stakeholders to find funding sources is an excellent approach. The SAN continuously works toward securing long-term funding to support future SWP efforts.

One major effort was the Schuylkill River Watershed Initiative (SAN 2004), which was funded in 2005 with a $1.15 million grant from the USEPA's Targeted Watershed Initiative Program. This program sought to demonstrate how water pollution can be managed on a watershed basis through the use of studies, demonstrations, and educational outreach activities. Projects selected through this program for the Schuylkill River were designed to demonstrate the effectiveness of collaborative networking in establishing priority projects that, when completed, would lead to improved water quality.

The Schuylkill River Watershed Initiative, administered by the Partnership for the Delaware Estuary, was designed to fund and implement more than 35 SWP projects. The SAN was integral in both determining the projects to be performed and for implementation of those projects. SAN Workgroup members, for example, collected and analyzed monitoring data, reviewed project outcomes and compared them to workgroup goals, and evaluated project success. SAN Education and Outreach Workgroup members, with support from other SAN participants, were responsible for communicating results and lessons learned from the Schuylkill River Watershed Initiative at state and national conferences and workshops. Through these SAN efforts, the Schuylkill River Watershed Initiative provided a model for other watersheds to move from assessment to protection and demonstrated a cooperative approach to maintaining coordinated actions under the Safe Drinking Water Act and Clean Water Act for a large watershed.

Philadelphia Water continues action plan development based on new information, analyses, and goals established for the program. Using the SWAs and SWPPs as templates, the program follows a cyclical pattern of continual assessment, goal-setting, implementation, and monitoring, with each program element informing the others. The Schuylkill River Restoration Fund (SRRF), funded through Exelon, Philadelphia Water, and other watershed stakeholders, picks up where the Schuylkill Watershed Initiative grant left off in terms of ensuring continued implementation of high-priority projects.

The SRRF provides grants to support environmental projects that improve and protect water quality in the watershed. Beginning in 2010, Philadelphia Water joined Exelon Generation's Limerick Generating Station as an annual contributor to the SRRF. The Partnership for the Delaware Estuary (PDE) became a contributor in 2011, Aqua Pennsylvania followed in 2012, and MOM's Organic Market in 2014. Coca Cola contributed in 2015. Government agencies, nonprofits, businesses, and other organizations with projects ready for implementation apply to the SRRF and are responsible for project execution, monitoring, and documentation. Members of the SAN serve as technical experts for grant recipient selection to ensure applicant projects will be beneficial to the Schuylkill River watershed. The grants are administered by the Schuylkill River Heritage Area (SRHA), which is managed by the nonprofit Schuylkill River Greenway Association.

The SRRF is the mechanism through which Philadelphia Water is able to contribute to projects that support WCP goals. The WCP focuses on the following three priority sources of *Cryptosporidium*: upstream wastewater discharge point sources, agricultural land-use runoff, and animal vectors (e.g., goose control), as well as education and outreach. PWD addresses *Cryptosporidium* in the watershed both by implementing SWPP initiatives and WCP-specific structural and nonstructural control measures in the watershed. The WCP control measures include supporting the installation of manure storage basins, stormwater management practices, and vegetated buffers on 10 farms throughout the Schuylkill River watershed; implementing a riparian buffer to deter animal vectors at a select site; and implementing a waterfowl management program at priority locations in Philadelphia. After the first three years of WCP implementation (2013–2015), Philadelphia Water has contributed to the installation of six manure storage basins on farms upstream of Philadelphia through the SRRF. Additionally, Philadelphia Water has a contract with the US Department of Agriculture to manage geese and other wildlife populations at its wastewater and water treatment plants and park properties in the city.

Additional considerations for the action plan are listed in the AWWA/ANSI C300 Standard as follows (AWWA 2014):

4.4.6   *Compliance with regulatory requirements.* The utility should determine and document relevant local, state or provincial, federal, or other source water protection regulations that apply to their utility and its source water protection area. The utilities shall comply with all applicable regulations for source water protection.

4.4.7   *Security planning and implementation.* The utility should have documentation that addresses security issues and describes, in detail, the response of personnel in the event of a security incident. Elements that address the protection of personnel and the water supply should be included. The applicable vulnerability assessment should be reviewed, and consideration must be given to access control and other relevant security issues.

4.4.8   *Emergency preparedness and response.* The utility should have documentation that describes emergency plans and provides specific directions to personnel in the event of an emergency. The program shall satisfy applicable regulatory requirements.

4.4.9   *Health and safety management.* The utility should document health and safety procedures that are designed to safeguard the employees and visitors engaged in operational activities pertaining to watershed management. The documentation may be specific to the source water(s) or part of a company-wide program.

Philadelphia Water programs that address the additional considerations for the action plan of the ANSI/AWWA (Requirement 4.4) are discussed below.

Requirement 4.4.6, Compliance with regulatory requirements. Philadelphia Water has consistently been an industry leader in not only compliance with relevant regulations but also in helping to lead numerous industry efforts in guiding development of important national regulations. As one related example, Philadelphia Water contracted a private consultant to work with them and the AWWA Office of Government Affairs to comment on USEPA's draft guidance manual for the Watershed Control Program component of the LT2-ESWTR. A revised draft of that guidance manual that was prepared as part of that effort was adopted nearly in full by USEPA. Philadelphia Water was one of the first utilities in the nation to develop a WCP in compliance with the regulation. In 2015, the program completed its third year of implementation for the five-year plan. Philadelphia Water submits a WCP annual status report to PADEP, as required by the regulation (PWD 2014, 2015a, 2016). The reports detail the progress made each year toward WCP goals and are available online at phillywatersheds.org. Additionally, Philadelphia Water became one of the first utilities to complete a Watershed Sanitary Survey as required by the LT2-ESWTR. The Watershed Sanitary Survey was submitted to the state at the end of 2015 and is also available online at www.phillywatersheds.org (PWD 2015b).

Requirement 4.4.7, Security planning and implementation. Philadelphia Water drinking water supply security revolves around a robust early warning system for advanced notice of potential water quality problems, an on-line monitoring system, a reservoir management standard operating procedure that outlines responses in the event of a security incident, a sampling and lab response plan in the event of contamination, and a microbial risk communication plan. Philadelphia Water also built a contaminant warning system throughout its distribution system in order to ensure timely detection and response and to mitigate health and economic consequences in the event of security breaches.

Requirement 4.4.8, Emergency preparedness and response. Philadelphia Water has developed an extensive emergency response plan and has dedicated personnel who are on-call 24 hours a day to respond to potential emergency incidents. The Industrial Waste Unit has staff trained to respond to and investigate all spills, accidents, and emergencies. At each WTP, staff have spill response manuals with specific watershed and plant information. Philadelphia Water intake pumping stations are equipped with floating booms and nets as well as spill-absorbent pillows to catch and absorb slicks of chemicals that float on the surface. Philadelphia Water also has a contract with a private company to provide emergency deployment and cleanup of spills at the raw water intakes. Philadelphia Water treatment plants are designed with raw water storage capacity, advanced hydraulic modeling capabilities, and a remotely operated distribution system, which provide the option of closing an intake for several days if necessary while minimizing service interruptions.

Requirement 4.4.9, Health and safety management. Philadelphia Water safety protocols are fully documented in standard operating procedures for personnel who conduct field activities such as water quality sampling and flow monitoring. The Philadelphia Water Safety Team is actively working to implement permanent changes in order to strengthen the Culture of Safety at Philadelphia Water. Since May 2015, there have been increases in the type and frequency of safety training courses, preshift safety talks, employee safety self-inspections, near miss/good catch reporting, safety refresher courses, employee safety surveys, a safety newsletter, recognition of safety accomplishments, and facility-specific incident tracking.

Program Implementation

Implementation of the action plan is the centerpiece of a successful SWP Program. Philadelphia Water has identified the following factors as critical to successful implementation of their SWP Program and projects:

- Stakeholder buy-in and support
- Funding
- Ability to address limited access or ownership land issues for much of the watershed
- Ability and capacity to implement projects successfully and build a track record of success
- Local stewardship and long-term sustainability
- Education and outreach to publicize the value of SWP projects to the local community and others both before and after project completion
- Monitoring of successes and evaluation of failures

The AWWA/ANSI G300 Standard requirements for program implementation of a SWP program are as follows (AWWA 2014):

Sec. 4.5   Program Implementation

Implementation of the action plan is the key to a successful source water protection program. Responses to unexpected challenges and barriers to implementation of the action plan items will also be assessed in determining compliance with this standard. It is expected that some projects may be led or conducted by stakeholder other than the water utility or utilities . . .

4.5.1   The utility shall, where appropriate, develop, promote, or implement a combination of voluntary and regulatory programs and sound practices such as

- Watershed planning
- Wellhead protection planning
- Land conservation
- Land use controls
- Contaminant source management
- Contingency planning
- Education and training
- Outreach and awareness programs
- Riparian buffers
- Green infrastructure and low-impact design standards
- Erosion and sediment control programs for construction projects
- Stormwater best management practices (BMPs)
- Agricultural best management incentives
- Watershed stewardship programs
- Responses to impacts from climate change and extreme events such as droughts and floods
- Security plan
- Health and safety plan
- Risk mitigation plan
- System vulnerability plan
- Operations plans

The Philadelphia Water SWP Program includes various regulatory, structural, and nonstructural controls to achieve effective protection and covers each category of management practices suggested in Sec. 4.5, Program Implementation, of the ANSI/AWWA (AWWA 2014). Philadelphia Water's efforts are quite diverse and include source identification studies, watershed monitoring, piloting of protection technologies, education programs, watershed partnerships, and a variety of implementation projects. Philadelphia Water has conducted several projects within its political jurisdiction (the City of Philadelphia). Although Philadelphia Water cannot directly conduct projects outside of their jurisdiction, which includes 98% of the Schuylkill River watershed land area, they have worked closely with the SAN to coordinate and provide technical information and assistance for numerous projects conducted further upstream in the watershed.

The Philadelphia Water SWPP works within the existing regulatory and political frameworks to address new potential sources of contamination to the SWP area. Within the City of Philadelphia, many of these new potential sources are addressed through Philadelphia Water's current authority or through coordination with other city agencies that have enforcement roles. Philadelphia Water conducts review and approval of new sewage facilities and stormwater control plans and coordinates with the Licenses and Inspections Unit, Planning Commission, and Department of Public Health on other related issues. This includes commenting and reviewing permit applications for new discharges and new construction or zoning changes. This input allows Philadelphia Water to suggest low-impact development techniques or additional protective measures and to positively impact the development of various sites.

Outside the City of Philadelphia, where Philadelphia Water does not have enforcement authority, they nonetheless provide review and comment on permits for discharges and construction to various federal, state, and local agencies. Pennsylvania is a commonwealth, which gives individual boroughs and townships the authority to approve development of areas without county oversight, thus making control of new potential sources more challenging. Philadelphia Water has worked with various county planning commissions to facilitate the adoption of model stormwater control ordinances, riparian buffer protection ordinances, agricultural preservation zoning, and conservation zoning in the critical areas of more than 200 townships in the watershed. Also, Philadelphia Water works with county conservation districts upstream of Philadelphia that review controls for all proposed development upstream. In addition, Philadelphia Water supports initiatives by local organizations upstream that are working to prevent new potential sources of contamination. It has been Philadelphia Water's experience that the voters and organizations within a local community tend to have greater and more immediate impact on decisions by local officials than other means.

Philadelphia Water's SWPP uses the following tools to implement improvement projects outside of their jurisdiction:

- Combination of the Safe Drinking Water Act and Clean Water Act regulatory frameworks and initiatives in a cohesive and productive manner
- A strong emphasis on the development of partnerships with upstream communities to achieve common goals while leveraging outside funds
- The latest science and technology to focus limited resources and protection efforts on critical areas that will result in the most cost-effective benefit to water supplies
- Coordination and cooperation with other water suppliers in the region to pool resources and efforts toward common goals and reduce overlapping efforts

Through the various techniques discussed above, Philadelphia Water is able to encourage the direction of activities upstream of its intakes without any regulatory authority, control, or jurisdiction. This allows Philadelphia Water to better leverage City resources and maximize the benefits of upstream protection efforts.

Philadelphia Water and its partners in the SAN and elsewhere have developed and implemented an impressive number of programs and projects. Information on implementation of SAN projects can be found at http://www.schuylkillwaters.org/projects.cfm and in the annually published progress reports, also available online at www.schuylkillwaters.org/san_publications.cfm. Additionally, the SRRF provides an ongoing means for Philadelphia Water to support implementation of projects upstream that support goals outlined in both its SWPP and WCP by leveraging funding and expertise from SAN partners and other watershed stakeholders.

Since the SRRF was established in 2006, more than $2.5 million has been collected, and grants have been awarded to 73 projects. In 2011, Land Protection Transaction grants were introduced as a part of the SRRF. This allows matching grants to be awarded up to $4,000 to cover transaction expenses for each conservation easement or other land protection transactions. SRRF and Land Protection Transaction grant guidelines are available on the SRHA website at www.schuylkillriver.org/Grant_Information.aspx. Grant recipients from the SRRF are selected by

Figure CS1-4 Agricultural best management practices, Including a manure storage basin and animal crossing, implemented at a farm in the Schuylkill River Watershed

a committee comprised of representatives from Exelon, DRBC, PWD, USEPA, PADEP, PDE, SRHA, and SAN. Projects address contamination from abandoned mine drainage, agriculture, and stormwater runoff. A number of projects have supported Philadelphia Water's WCP and compliance with LT2-ESWTR. Figure CS1-4 shows an example of agricultural stormwater best management project implementation. Table CS1-2 lists grant recipients, their projects, and the awarded amounts. The grants awarded to projects through the SRRF are leveraged many times over. Applicants are encouraged to secure funding from private, nongovernmental sources; the grant selection committee gives priority in the ranking process to projects with high levels of match. SRRF grant applicants, which are required to have a minimum 25% match, frequently secure significantly more in the form of cash and in-kind from federal, state, and local grant programs; local water utilities; municipalities; conservation districts; private businesses; nonprofits; school districts; community groups; and other sources.

Table CS1-2 Schuylkill River Restoration Fund Grant Recipients for 2006–2015

Grant Recipient	Grant Amount	Project
2015		
Berks County Conservancy	$47,250	Agricultural improvements at one Berks County farm
Berks County Conservancy	$81,663	Agricultural improvements at two Berks County farms
The Dobson School	$37,710	Woodland walk and native meadow for stormwater management and outdoor classroom space
Montgomery County Conservation District	$30,000	Stormwater basin retrofit
Schuylkill Headwaters	$40,000	Removal of coal refuse and restoration of natural flood plain of Schuylkill River, Schuylkill County
West Philadelphia Coalition	$26,000	Stormwater project to transform barren school yard at Lea Elementary School in West Philadelphia
Berks County Conservancy	$4,000	Conservation easement for Miller Woodland property
Berks County Conservancy	$4,000	Purchase of Gehris property for conservation easement
Natural Lands Trust	$4,000	Purchase of Yoder tract for riparian easement
2014		
Berks County Conservancy	$80,577	Agricultural improvements at one Berks County farm
Berks County Conservancy	$49,350	Agricultural improvements at one Berks County farm
Berks County Conservation District	$80,577	Agricultural improvements at one Berks County farm
Borough of Pottstown	$30,000	Water quality infiltration filters, Manatawny Creek and Schuylkill River
Cook Wissahickon School	$37,961	Meadow restoration project (phase 2), Schuylkill River
Schuylkill Headwaters Association	$55,000	Stormwater BMPs on Blue Mountain School District campus
Lower Frederick Township	$4,000	Stonehill Greenway property purchase
2013		
Berks County Conservancy	$99,750	Agricultural improvements at two Berks County farms
Montgomery County Conservation District	$20,000	E. Norriton Middle School stormwater BMPs
Partnership for the Delaware Estuary	$20,821	Schuylkill Action Students–stormwater BMPs at three schools
Schuylkill Headwaters Association	$63,000	Limestone for Mary D abandoned mine drainage facility
Valley Forge Trout Unlimited	$25,000	Crabby Creek infiltration trench
Berks County Conservancy	$68,250	Agricultural improvements at one Berks County farm
Roxborough Development Corporation	$50,000	Leverington Parklet Project
Natural Lands Trust	$4,000	Killian Woods property purchase
Natural Lands Trust	$4,000	Corbett property purchase
Wissahickon Valley Watershed Association	$4,000	Pizeak property open space preservation
2012		
Berks County Conservancy	$99,750	Agricultural improvements at two Berks County farms

Meadowood Retirement Community	$58,000	Stormwater basin retrofit, Skippack Creek
Pennsylvania Horticultural Society	$38,500	Hunsberger Woods stormwater project, Perkiomen Creek
Wissahickon Sustainability Council	$27,065	Native meadow, Schuylkill River
Berks County Conservancy	$4,000	Russ property conservation easement, Schuylkill County
Berks County Conservancy	$4,000	Gehman property conservation easement, Berks County
Montgomery County Lands Trust	$4,000	Rogers property easement, Montgomery County
Natural Lands Trust	$4,000	Green Hills property easement, Berks County
2011		
Berks County Conservancy	$97,755	Agricultural improvements at three Berks County farms
Greening Greenfield	$50,000	Green roof installation at Albert M. Greenfield School
Maidencreek Township	$30,000	Willow Creek habitat restoration
Montgomery County Conservation District	$55,000	Schuylkill Action Students–Stony Creek restoration
The Schuylkill Project	$60,000	Shawmont restoration project
Berks County Conservancy	$4,000	Oley Hills land protection project, Berks County
East Pikeland Township	$4,000	George Lang property
2010		
Berks County Conservancy	$52,500	Martin Farm agricultural improvements
Lower Providence Township	$40,000	Five stormwater basin retrofits, Perkiomen Creek watershed
Schuylkill Headwaters Association	$80,000	Glendower Breach abandoned mine drainage project
Greening Greenfield	$50,000	Schoolyard stormwater project
2009		
Berks County Conservancy	$75,000	Agricultural improvements at one Berks County farm
Schuylkill Headwaters Association	$100,000	Wheeler Run acid mine drainage project
2008		
Berks County Conservancy	$15,500	Agricultural improvements at one Berks County farm
Montgomery County Lands Trust	$25,000	Stormwater infiltration system, Scioto Creek
Pennsylvania Environmental Council	$97,400	Stormwater detention basin retrofit demonstration project, Wissahickon Creek
2007		
Schuylkill Headwaters Association	$61,141	Bell Colliery, Reevesdale, and Otto acid mine drainage projects
Berks County Conservancy	$98,500	Agricultural improvements at two Berks County farms
2006		
Berks County Conservancy	$50,931	Agricultural improvements at two Berks County farms
Berks County Conservation District	$48,750	Agricultural improvements at one Berks County farm
Perkiomen Watershed Conservancy	$20,175	Stormwater remediation, Unami Creek
TOTAL	**$2,325,876**	

Abbreviation: BMP, best management practice.

The following are examples of ongoing SWP projects for the Schuylkill River watershed.

Cryptosporidium Monitoring Research

Figure CS1-5 Philadelphia Water identified geese as sectors of *Cryptosporidium* through source tracking research

For a decade, Philadelphia Water has worked in collaboration with Lehigh University to better understand the occurrence, sources, and vectors of *Cryptosporidium* in the Schuylkill River watershed. This research informs Philadelphia Water WCP program goals. Sampling programs are designed to answer research questions and improve and expand methods for field sample collection and laboratory analysis of *Cryptosporidium*. Through this research, Lehigh University identified geese as vectors of *Cryptosporidium* (Figure CS1-5), and Philadelphia Water has implemented programs to deter these animals from sensitive SWP areas.

Friday, July 03
Current RiverCast:

RED

Terms of Use

Water Temp: **69 °F | 20 °C**
River Flow: **5880 cfs**

Figure CS1-6 Example of Philly RiverCast rating, river temperature and flow as displayed on the website and updated hourly

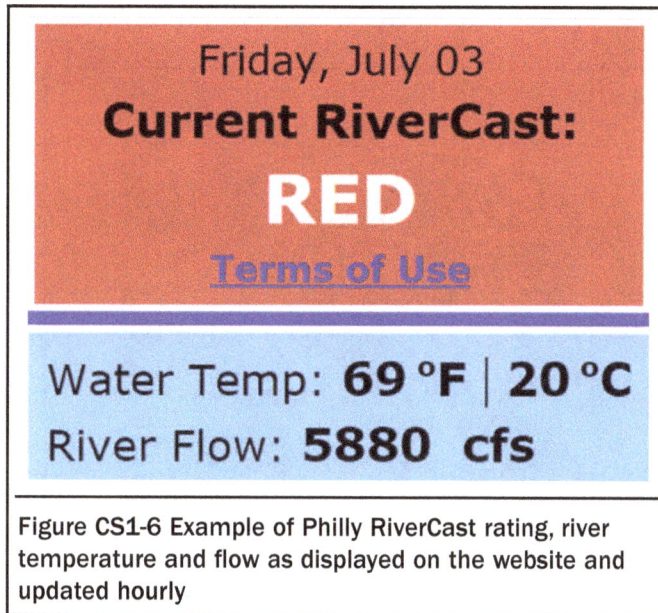

Philly RiverCast

Philly RiverCast is a web-based hourly forecast of bacteria levels that guides recreational users of the Schuylkill River from the Fairmount Dam to Flat Rock Dam in Manayunk (www.phillyrivercast.org). After extensive data analysis, the PWD developed an algorithm that incorporates real-time measurements of precipitation, flow, and turbidity to trigger a red, yellow, or green rating based on USEPA recreational guidelines. The system was designed to be a public health tool and, as such, maximizes accuracy while minimizing the occurrence of false positives, or indicating that the water quality is better than it is likely to be.

Green RiverCast ratings suggest the water quality is suitable for all recreational activity. A yellow rating indicates that the water quality may not be suitable for primary or direct recreational contact (e.g., water skiing, kayaking, swimming, wading), and a red rating indicates that the water quality is not suitable for primary recreational contact and may not be suitable for secondary or indirect recreational contact (e.g., rowing, fishing, boating). At the time of this publication, the website has received more than 800,000 hits.

Education and Outreach Stormwater Guides

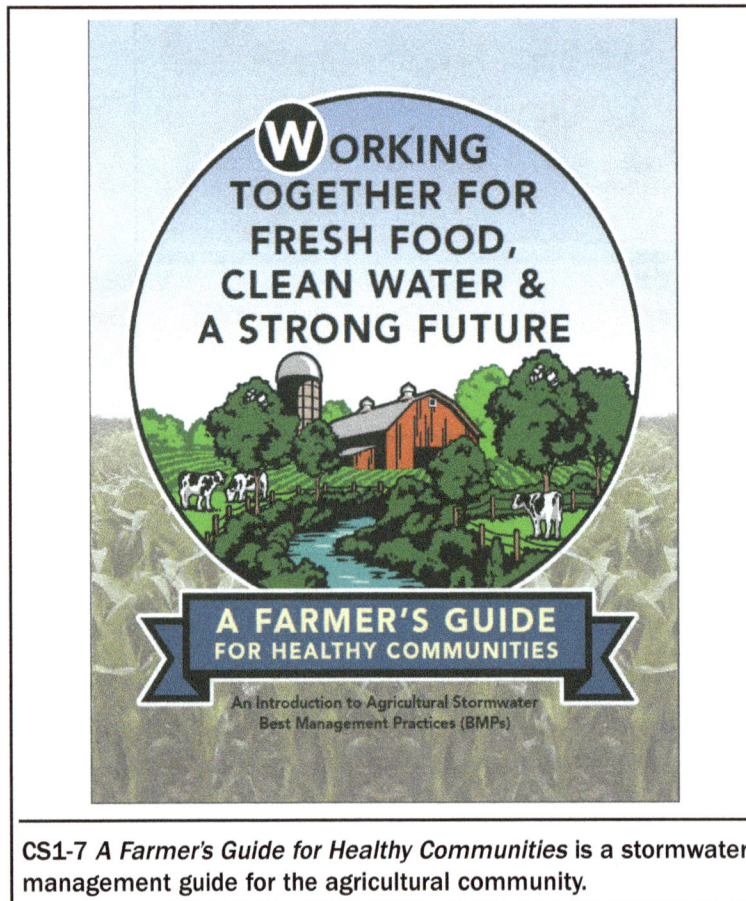

CS1-7 *A Farmer's Guide for Healthy Communities* is a stormwater management guide for the agricultural community.

Through the SAN Education and Outreach Workgroup, Philadelphia Water supports education of the public on how their actions affect the Schuylkill River watershed, raising awareness for problems and their solutions and gaining support for BMP projects. The SAN and Philadelphia Water have published numerous materials relating to stormwater management for schools, home owners, and commercial property owners. The more recent guide, *A Farmer's Guide for Healthy Communities,* was completed in 2014 for farmers and agricultural landowners to emphasize the benefits of stormwater and manure management for both their business and the watershed.

Schuylkill Watershed Priority Lands Strategy

The Schuylkill Watershed Priority Lands Strategy is a geographic information system (GIS) modeling tool used to identify areas in the watershed that should be preserved for ecological and drinking SWP purposes. The concept for the strategy originated from discussions within the SAN Watershed Land Collaborative

Workgroup and the following core partners: Natural Lands Trust (NLT), Philadelphia Water, and the Delaware Valley Regional Planning Commission (DVRPC). The project (Figure CS1-6) received a Growing Greener grant from the PADEP in the fall of 2005, began in the winter of 2006, and was completed in the spring of 2007.

A Combined Priority Map was created by identifying areas important for ecological conservation and drinking water supply protection. NLT developed SmartConservation, a science-based method that is used to prioritize ecologically valuable landscapes. Philadelphia Water developed a model that ranks land not already developed or protected according to importance for surface and groundwater protection. The Combined Priority Lands map is a compilation of SmartConservation and the Drinking Water Source Protection model that identifies priority lands of both ecological and drinking water supply importance. In the Combined Priority Lands map, 172,000 acres (14%) of the Schuylkill River watershed are identified as a high-priority resource to ecological conservation and drinking water supply protection.

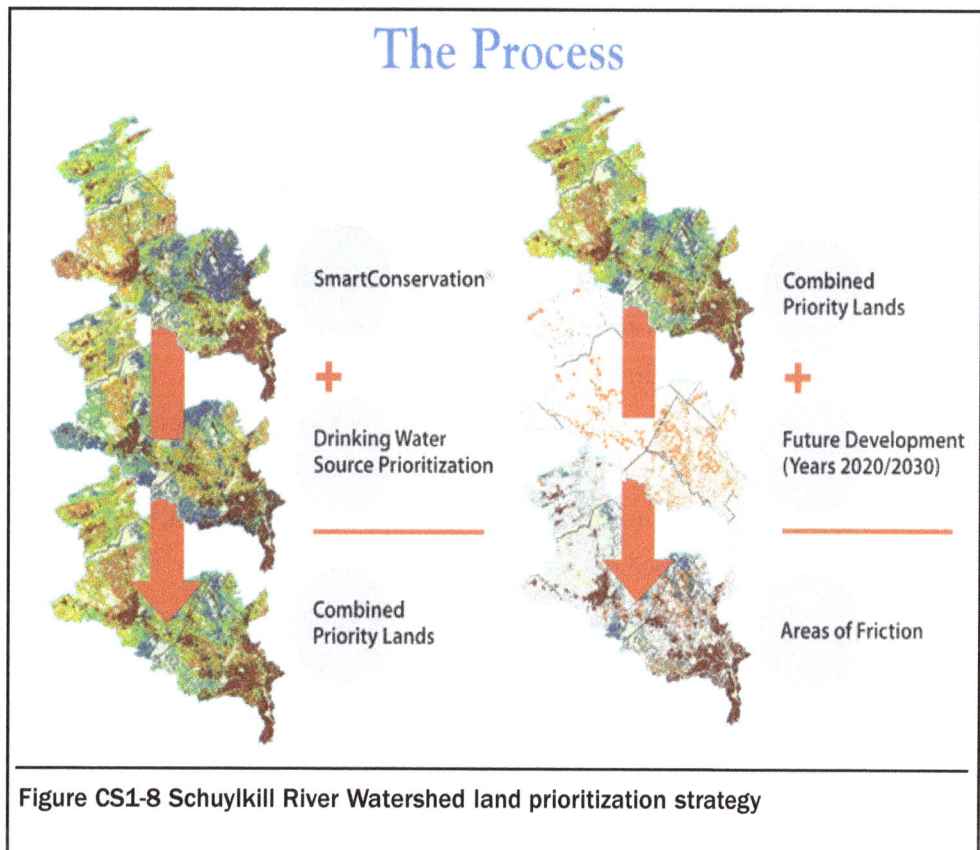

Figure CS1-8 Schuylkill River Watershed land prioritization strategy

A model for development potential was created by combining the DVRPC Region Growth Model, Berks County Growth Model, and the Schuylkill County Growth Model. It identifies 65,600 acres of new development from growth over the next 10 to 20 years. The DVRPC also created a Land Consumption map that includes the relative number of acres by municipality of new development forecasted for 2030. The Combined Priority Land maps and Future Development maps were combined to create an Areas of Friction map. Identification of areas of friction will help avoid development in areas that are important for healthy ecosystems and clean drinking water. Of the total estimated acres of new development, only 10% are considered areas of friction. The most prominent areas of friction are located in the Pickering Creek watershed, southern and eastern portions of Berks County, and the Upper Perkiomen Valley in Montgomery County.

Select examples of ongoing SWP projects in the Delaware River watershed are described below.

Delaware Valley Early Warning System

The EWS is an integrated monitoring, notification, and communication network designed to identify and provide advance warning about source water contamination events (i.e., intentional or accidental discharges to the water supply or naturally occurring changes in water quality). Developed in partnership with numerous water suppliers and government agencies in the Schuylkill River and Delaware River watersheds, this system includes integrated telephone and Internet capabilities for contamination event notification, a water quality monitoring network for real-time and historical monitoring data, and database and data management capabilities that are fully integrated into a user-friendly website. The website contains additional analytical tools such as spill routing and estimates of spill time-of-travel to the various water intakes. Philadelphia Water obtained the initial funding, approached the other water suppliers on the river, led organizational meetings, and developed the web-based communication system for the early warning system. In combination with Philadelphia Water's emergency response plans, the Delaware Valley Early Warning System provides a means to protect both public water supply and industrial intakes from sudden, major contamination episodes such as large chemical or oil spills. The website mapping tools were upgraded with ArcGIS technology in 2013, and a sophisticated tidal spill trajectory tool was incorporated into the system in 2014 (Figure CS1-9). Philadelphia Water was awarded the 2015 Governor's Award for Environmental Excellence for this tool. At the time of this publication, the EWS user base consisted of more than 300 individuals from approximately 50 organizations.

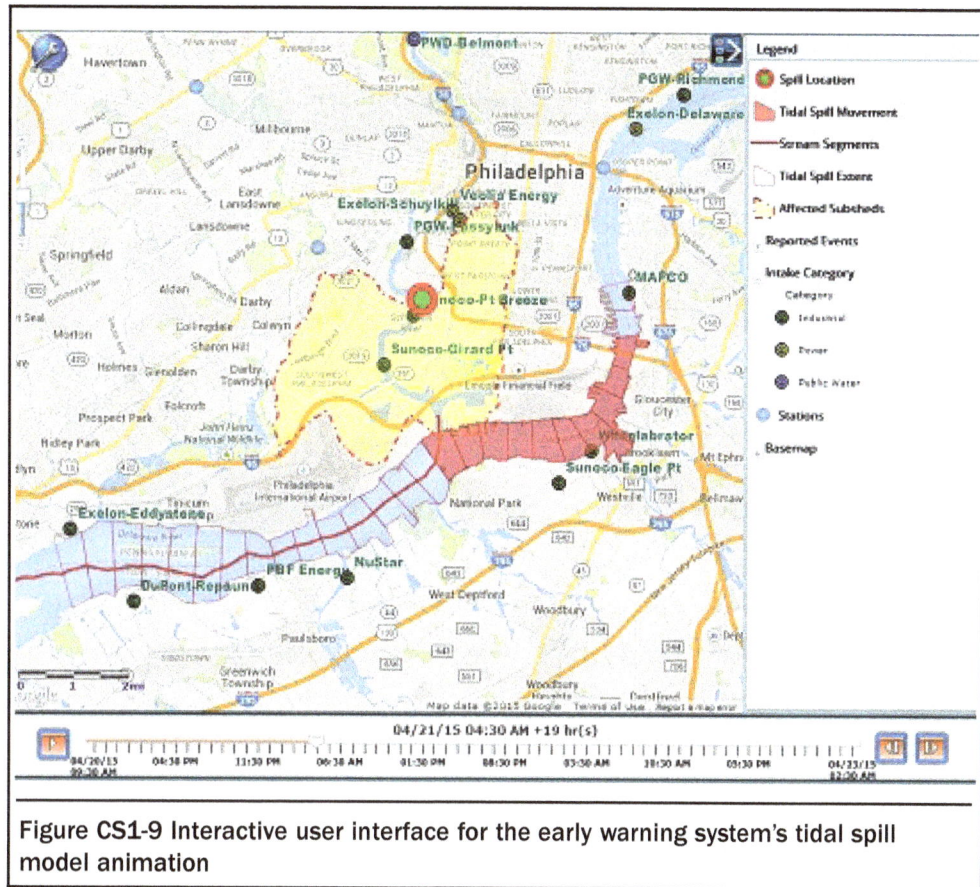

Figure CS1-9 Interactive user interface for the early warning system's tidal spill model animation

Delaware River Flow Policy

Philadelphia Water uses operational analysis and simulation of integrated systems (OASIS) software (Hydrologics Inc.) to simulate watershed management policy alternatives in the Schuylkill River and Delaware River watersheds, including all streams and reservoirs. Water supply modeling is a technique Philadelphia Water uses to analyze how current policies and alternative policies may influence reservoir storage availability and streamflow during seasonally wet and dry conditions, moderate to severe drought, and emergencies.

States within the basin use an OASIS model of the Delaware River Basin to evaluate how available water resources may meet competing flooding, fisheries, water supply, drought, power, and water quality interests. The software allows users to replicate the watershed of interest by simulating historical observed streamflow at multiple calculation points and reservoirs. The model then allows reservoir release policies, demand, discharges, and other water resource features to be adjusted in order to study how each policy configuration would act on historical conditions. This allows for side-by-side comparison of historical conditions and how those conditions would react to different management policies. Philadelphia Water serves as a technical advisor to the State of Pennsylvania in ongoing negotiations among the parties to the 1954 Supreme Court Decree to manage streamflow in order to support competing water needs in the Delaware River Basin.

Evaluation and Revision

SWPPs need to be periodically evaluated and revised in response to changes in the source water area, new information, performance of implemented programs, and similar changes. The AWWA/ANSI G300 Standard requirements for Sec. 4.6, Evaluation and Revision, of a SWPP are as follows (AWWA 2014):

Sec. 4.6   Evaluation and Revision

Source water protection programs shall be periodically evaluated and revised in response to changes in the area of source water delineation, new data or information, new regulatory initiatives, changes in local priorities, actual performance of implemented programs, and so forth.

4.6.1   Evaluation procedures. The source water protection program shall include provisions for periodically reviewing and, if necessary, modifying the utility's source water protection vision, characterization, goals, and implementation elements. This adaptive management approach (as a step of the process) is intended to measure the accomplishment or completion of projects, programs, and activities identified in the action plan. It also aims to identify gaps and shortcomings in the program for making future improvements.

In 2005, Philadelphia Water hired an outside consultant to perform an independent benchmarking evaluation of its SWPP. This evaluation included comparison of the SWPP to published guidelines and industry standards, as well as benchmarking the Philadelphia Water program against advanced SWPPs developed by other US drinking water utilities. The intended objective was to identify potential program deficiencies and develop recommendations of new ideas for program activities to help improve the program's effectiveness. The program was evaluated again in 2008 in addition to a more extensive assessment of the source water programs of other high-profile cities in order to evaluate and benchmark

Philadelphia Water's own program. In 2014, the SWPP was identified as a core mission of Philadelphia Water's Strategic Plan. As of 2015, the Philadelphia Water Office of Watersheds and SWPP are more than 15 years old. The goals and activities of these programs are under continual evaluation.

Conclusions and Lessons Learned

The Philadelphia Water SWPP serves as a model for drinking water utilities across the nation. Philadelphia Water has enthusiastically adopted for itself the role of the region's "watershed champion," accepting the challenge of implementing a comprehensive, multibarrier approach to water resource management. Their innovative vision and mission are supported by a commitment from the utility to work toward achieving those objectives.

The Philadelphia Water program includes each key component generally considered necessary for an effective SWPP. For example:

- A clear vision articulated by Philadelphia Water and continually supported with sufficient resources to accomplish and maintain the vision
- A SWP area delineation that satisfied both state and federal requirements
- Source water assessments that go well beyond that required by the Safe Drinking Water Act regulations
- A SWPP and program developed in collaboration with a stakeholder team that is representative of the community
- An established management program to effectively control known and/or potential existing and new sources of contamination into the SWP area
- An emergency plan that allows for efficient and effective emergency response to manage contamination events that may threaten the water supply within the SWP area
- Ongoing community education, outreach, and involvement
- A SWPP that engages county, state, and federal authorities who regulate potential sources of contamination within the delineated SWP area
- A SWPP with a full-time staff responsible for carrying out the effort

Philadelphia Water SWP efforts are particularly challenging in that their raw water intakes are located at the bottom of two major rivers and most of the land area in these watersheds is outside of Philadelphia County, making any impact on local land-use/development decisions quite difficult. Nonetheless, Philadelphia Water has risen to this challenge through substantial outreach to local governments and many other stakeholders. In many instances, Philadelphia Water has turned this disadvantage into an asset by reaching out to numerous partners in order to share and leverage resources to address various common water resource issues and by developing a sustainable and coordinated framework (e.g., the SAN) in order to achieve true environmental outcomes and benefits. In this manner, Philadelphia Water can work within and support existing frameworks and initiatives to influence upstream solutions without extensive resource investments.

As successful as the SAN has been in coordinating upstream activity, Philadelphia Water has recognized that development of a policy agenda at the state level is key for widespread change. Also, the first 10 years of the Philadelphia Water SWPP focused almost exclusively on water quality. An equal emphasis on water quantity and the impact of reservoirs on water supply must also play a role, and Philadelphia Water has recently been focusing on those areas.

In the course of developing the Office of Watersheds and SWPP, various changes in departmental policies and approaches were made within Philadelphia Water. The effort took a substantial commitment by Philadelphia Water's leadership and staff and involved enhanced coordination of funding and resources.

Keys to the success of Philadelphia Water's SWP Program include the following:

- An in-depth understanding of the watershed, existing threats, and potential threats to the integrity of the water supply
- Established and long-standing watershed stakeholder partnerships for coordination and collaboration
- Knowledge of regional watershed protection priorities
- Organized watershed partnership groups to implement protection efforts
- A sustainable source of funding for protection efforts

Some of the important lessons Philadelphia Water has learned include the following:

- Scientifically sound data is needed to support identified potential water quality and quantity issues.
- Customer communications and stakeholder involvement are key to getting people to understand the issues and to take the necessary steps to address them.
- Significant resources are needed to gain a stakeholder consensus and resolve issues.
- Implementation requires significant financial resources and people.

- Continued program success relies heavily on the effective communication of results and program progress.

References

American Water Works Association. 2014. ANSI/AWWA Management Standard G300 - Source Water Protection. Denver, Colo.: AWWA.

Maimone, M., Crockett, C., and Cesanek, W. 2007. "PhillyRiverCast: A Real-Time Bacteria Forecasting Model and Web Application for the Schuylkill River." *Journal of Water Resources Planning and Management*, 133(6), 542-549.

Delaware River Basin Source Water Collaborative (DRBSWC). 2016. <http:// www.delawarebasindrinkingwater.org/>

Philadelphia Water Department (PWD). 2002a. *Belmont & Queen Lane Treatment Plants (PWSID #1510001) Source Water Assessment Report*. Philadelphia, PA: PWD. <http://www.phillywatersheds.org/doc/Schuylkill_SWA_lowres.pdf>

———. 2002b. *Baxter Water Treatment Plant (PWSID #1510001) Source Water Assessment Report*. Philadelphia, PA: PWD. <http://www.phillywatersheds .org/doc/Delaware_SWA.pdf>

———. 2006. *The Schuylkill River Watershed Source Water Protection Plan (Belmont & Queen Lane Surface Water Intakes)*. Philadelphia, PA : PWD. 218 pp. <http://www.phillywatersheds.org/doc/Schuylkill_SWPP_2006.pdf>

———. 2007. *The Delaware River Watershed Source Water Protection Plan (Baxter Water Treatment Plant Surface Water Intake)*. Philadelphia, PA : PWD, 195 pp. <http://www.phillywatersheds.org/doc/Delaware_SWPP_2007.pdf>

———. 2009. *Philadelphia Combined Sewer Overflow Long Term Control Plan Update*. "Section 1: Introduction and Background." Philadelphia, PA: PWD, 33 pp.

———. 2010. *Schuylkill River Hydrology and Consumptive Use*. Philadelphia, PA: PWD, 75 pp.

———. 2011a. *Amended Green City Clean Waters: The City of Philadelphia's Program for Combined Sewer Overflow Control Program Summary*. Philadelphia, PA: PWD, 48 pp.

———. 2011b. *Long Term 2 Enhanced Surface Water Treatment Rule Watershed Control Program Plan Queen Lane Drinking Water Treatment Plant Schuylkill River, Philadelphia, PA*. Philadelphia, PA: PWD, 116 pp. http:// phillywatersheds.org/doc/Sourcewater/PWD_Watershed_Control_Plan_ final.pdf

————. 2014. *2013 Watershed Control Plan Annual Status Report*. Publication. www. phillywatersheds.org/what_were_doing/documents_and_data/watershed _plans_reports

————. 2015a. *2014 Watershed Control Plan Annual Status Report*. Publication. www.phillywatersheds.org/what_were_doing/documents_and_data/ watershed_plans_reports

————. 2015b. *2015 Watershed Sanitary Survey*. Publication. www.phillywatersheds. org/what_were_doing/documents_and_data/watershed_plans_reports

————. 2015c. Philadelphia Water Strategic Plan. Philadelphia, PA: PWD.

————. 2016. *2015 Watershed Control Plan Annual Status Report*. Publication. www. phillywatersheds.org/what_were_doing/documents_and_data/watershed _plans_reports

Schuylkill Action Network (SAN). 2004. *The Schuylkill River Watershed Initiative: Workplan*. Philadelphia, PA: PWD, 30 pp.

————. 2013. *2003–2013 Progress Report*. Wilmington, DE: Partnership for the Delaware Estuary, 17 pp.

Case Study: Central Arkansas Water, Little Rock, Arkansas

Central Arkansas Water (CAW) provides water to more than 400,000 individuals throughout the central Arkansas area, including the cities of Little Rock and North Little Rock. CAW obtains its source water from two high-quality and well-protected surface water supplies, Lake Maumelle and Lake Winona. Lake Winona is a 1.9-square mile reservoir built in the 1930s that provides approximately 40% of the daily system-wide demand for CAW. The lake has a 43-square mile (27,520-acre) watershed that is almost completely forested and is located in the Ouachita National Forest. Lake Maumelle is a 13.9-square mile reservoir that provides approximately 60% of daily system-wide demand. The lake was developed in the 1950s as a second raw water supply to Lake Winona. Approximately 91% of Lake Maumelle watershed's 88,000 acres (137.5 square miles) is forested, with 53% privately owned and subject to potential development.

This case study focuses primarily on the Lake Maumelle watershed, as the Lake Winona watershed does not face the same development pressures as Maumelle's watershed. A total of 100% of the Winona watershed is either owned by CAW or managed by the US Forest Service and is afforded an additional layer of protection by virtue of being located in the Ouachita National Forest. As a result, a greater amount of resources has been invested in the development, implementation, and modification of source water protection (SWP) plans for Lake Maumelle. However, many of the watershed protection efforts discussed are also implemented in the Lake Winona watershed.

Lake Maumelle was built in the 1950s at an estimated cost of $34 million (in 2006 dollars). It would cost far more today to replace the high-quality water supplied by the lake. Lake Maumelle is a human-made reservoir that was designed and created to operate solely as a drinking water supply. CAW owns and operates Lake Maumelle as one of two principle water supplies for 15 cities and communities in the region. The lake and its watershed are critical to the quality of life and economy of the Central Arkansas region. It serves more than 400,000 people today, with a projection of more than 575,000 people by 2050. Data collected by the Arkansas Department of Environmental Quality, Central Arkansas Water, and the US Geological Survey (USGS) demonstrate that Lake Maumelle is one of the highest-quality water supplies in Arkansas and the southeastern United States. A primary reason for its high quality is that the Lake Maumelle watershed has remained substantially undeveloped.

Concerns about the degradation of water quality in Lake Maumelle date back to as early as 1985, when a study authored by researchers at the University of Arkansas at Little Rock predicted developmental pressures from the expansion of Little Rock would reach the Lake Maumelle watershed in 25 years and advocated for proactive protection measures. In 1988, a second study conducted by the Benham Group recommended long-term water quality monitoring of the lake and a watershed protection program. In response to this recommendation, in 1989 the Little Rock Municipal Water Works (LRMWW) and the USGS partnered on a long-term program to monitor water quality. This program provides data for future reservoir protection and modeling and for monitoring the impact of land-use changes on water quality and sediment deposition. LRMWW (CAW's predecessor) adopted its first watershed protection program in 1992, in which it identified lands for acquisition close to the water intake needed to provide greater water quality protections.

In 2000, a study by the University of Arkansas at Little Rock inspired the cities of Little Rock and North Little Rock to make a major change in their relationship by moving past geographical differences and corporate interests to benefit the entire customer base and surrounding area. The result was a decision by the cities' governing bodies and water commissions to merge LRMWW and the North Little Rock Water Department into Central Arkansas Water.

The continuing westward expansion of Little Rock along with multiple proposed developments in the early 2000s in close proximity to the Lake Maumelle intake compelled CAW to thoroughly evaluate the best way to safeguard the long-term protection of water quality in the lake. In 2004, CAW convened the Task Group for Watershed Management to evaluate CAW's watershed protection program and determine if development near the Lake Maumelle intake would degrade water quality.

The CAW Board approved the task group's recommendations in 2005, which included development of a comprehensive, science-based watershed management plan. CAW contracted with Tetra Tech and began development of the Lake Maumelle Watershed Management Plan (the Plan), which was adopted by the CAW Board in 2007. The Plan continues to guide implementation of the utility's watershed management efforts.

Utility leadership, key stakeholders, and community leaders support CAW's proactive SWP efforts. Rather than address watershed protection on a case-by-case basis in the courts, CAW and its community stakeholders have sought to develop and implement a robust watershed protection plan that addresses a wide range of causes and sources of potential impairment to the lake and its tributaries. CAW's watershed management efforts consist of land acquisition, land-use planning and regulation in collaboration with Pulaski County, water quality monitoring and assessment activities, other pollution control and management measures, and public education efforts.

Source Water Protection Program Vision

Prior to adoption of the Plan in 2007, CAW had not explicitly stated a vision for SWP. The CAW Board of Commissioners (the Board) adopted resolutions in 1992, 1998, and 2003 to establish plans to continue protection of the Lake Maumelle watershed. When the Board finalized and adopted the Plan in 2007, the Board established a clear vision for watershed protection where the utility and community would strive to "maintain [a] long-term, abundant supply of high quality drinking water for present needs and continuing growth of the community . . . and [provide] an equitable sharing of costs and benefits for protecting Lake Maumelle." CAW staff strives to meet the challenges of this vision and implement its core values.

Demonstration of the commitment of CAW for their SWP efforts is evident through the many initiatives that have been implemented, the funding that has been provided (including through a special designated watershed protection fee of $0.45 on customers' monthly bills), as well as through the creation and hiring of three staff positions related to SWP (described later). CAW has also proactively involved numerous stakeholders at various stages of the SWP planning and implementation process.

Characterization of Source Water and Source Water Protection Area

The Arkansas Department of Health conducted a source water assessment for LRMWW in 2000 as part of its assessment efforts for all Arkansas utilities. Since that time, CAW has invested heavily in the characterization of its source water and SWP areas through internal data collection and analysis and through collaborative partnerships with the USGS, state and local agencies, and independent consultants.

CAW has developed a robust geographic information system (GIS)–based watershed database that provides links to a wide range of aerial imagery, land-use data, water quality monitoring data, environmental permits, forest resource data, and other key natural or human-made features in the Lake Maumelle watershed. Information in the database includes high-resolution color and infrared photos for the Lake Maumelle watershed. With these resources, the utility is able to review and compare aerial imagery from 2002 to those from the present in order to identify key land changes. CAW has also used existing contracts to take aerial photographs of nearby lakes following major storm events to determine which tributaries are delivering the most turbid water to Lake Maumelle. Recently, CAW added the use of satellite imagery to its suite of land change and response tools and relies on internally developed and national land-cover data to inform its decision-making processes.

CAW populates its watershed GIS database through a wide range of partnerships. Since 1985, a large volume of land-use and water quality data has been collected by CAW through projects with the USGS. The goal is to provide data for future reservoir protection and modeling as well as for monitoring the impact of land-use changes on water quality and sediment deposition. CAW continues to have a strong source water quality monitoring program in partnership with the USGS for both Lake Maumelle and Lake Winona, with an annual budget of approximately $425,000.

CAW has worked with state agencies and consultant foresters to identify forest stand types in the Lake Maumelle watershed. These data provide a basis for CAW's forest management efforts by identifying the current condition of forests and allowing CAW to rely on expert recommendations for progression from current conditions to the desired future conditions of forested areas. Ultimately, knowledge of the current conditions in the watershed's forested areas allows CAW to select the preferred management tools and techniques for moving toward the future conditions that will optimize water quality and quantity provided by the forested landscape.

Source Water Protection Goals

To guide development of the Plan, Tetra Tech invited a wide range of stakeholders from across the region to join a 22-member Policy Advisory Council (PAC). Through a consensus-based process, the PAC adopted and prioritized nine general goals for the Plan. The goals, in order of priority, were to

- Minimize risk to public health
- Minimize impacts on watershed property owners and residents
- Minimize water supply taste, odor, and color problems
- Minimize impact on the water supply intake and water treatment operations
- Minimize rate increases
- Minimize loss of reservoir supply storage capacity from sedimentation
- Minimize risk of impairment to tributary streams in the watershed for stream and lake protection
- Allow limited recreation that reflects environmentally sound stewardship of the lake
- Meet other community values

The PAC also worked with consulting engineers and a technical advisory committee of scientists, modelers, and other technical resource advisors to establish water quality indicators and associated numeric targets that would allow the PAC to determine if the specific action items included in the Plan met the goals. The water quality indicators used in development of the Plan were chlorophyll-*a* concentrations, total organic carbon concentrations, lake turbidity, and fecal coliform bacteria. The use of numeric targets for these water quality indicators provides a benchmark that guides CAW in evaluating changes over time in the Lake Maumelle watershed and in adapting and modifying its management actions as necessary.

To predict how lake water quality would respond to various management alternatives, Tetra Tech developed watershed and lake models for development of the Plan. A baseline analysis was performed to compare existing watershed and lake conditions to two potential future build-out scenarios. The baseline analysis showed that without implementation of a wide range of action items, the Plan would not meet the numerical targets for the key water quality indicators and, therefore, would not achieve the goals established by the PAC.

Action Plan

As discussed previously, the Plan, which was adopted by the CAW Board in 2007, continues to guide implementation of the utility's watershed management efforts to this day.

The Plan specifically identifies a wide range of action items for CAW to pursue in order to meet the Plan's goals. The PAC developed the final list of action items included in the Plan using technical and public input, sound science and engineering, and key feasibility evaluations. The PAC solicited technical and public input on each management action, evaluated the impact of a given action on the numerical water quality targets established to accomplish the Plan's goals, and modified the management actions based on public and scientific input. Ultimately, compromise and consensus were required to finalize the list of management actions included in the Plan.

The Plan clearly states that "no single management option can meet all of the [goals of the Plan]; therefore a combination of methods and actions are needed." However, the Plan emphasizes that the largest threat to the lake is the conversion of forest to new housing development and associated wastewater. Therefore, the Plan recommends a strong focus on management of the impacts from new development.

The Plan further articulates specific strategies and action items in each of the following areas to ensure the long-term viability of the lake as a drinking water supply:

- Managing impacts from new development
- Acquiring conservation land
- Mitigating hazardous materials spills
- Improving and maintaining existing roads
- Continuing good lake management practices
- Continuing sound land management practices on CAW property and supporting plan implementation

- Maintaining good forest practices
- Encouraging good household practices
- Maintaining an adaptive approach and monitoring success

Implementation of the Action Plan

Implementation of the action plan is the centerpiece of a successful SWP Program. Prior to completion of the SWP Plan, CAW implemented a number of SWP measures including acquisition of high-priority land in the watershed. The Plan further identifies and prioritizes future implementation efforts for CAW and community stakeholders. The Plan also includes specific implementation components as action items, including hiring staff to implement the Plan as well as outlining the financial resources needed to accomplish the Plan's goals.

The Plan identifies the need to create three new CAW positions—a watershed protection manager to oversee the program, a watershed administrator to review permits and proposals for construction within the Lake Maumelle watershed, and a stewardship coordinator to coordinate various land management activities on CAW-owned property in the Lake Maumelle watershed. CAW created these three positions in the years that followed adoption of the Plan. Additionally, CAW reimburses Pulaski County for operational expenses and staff time associated with land-use planning and enforcement of watershed regulations in the Pulaski County portion of the Lake Maumelle watershed.

The Plan identifies immediate actions needed to address threats from new development as a critical, short-term action item. To address water quality threats from development, CAW partnered with Pulaski County to adopt and implement subdivision and zoning regulations to guide development in the Lake Maumelle watershed (Pulaski County currently has the highest potential for development in the watershed). CAW and Pulaski County have partnered to adopt regulations to establish site-specific nutrient and sediment loading limits, require that conservation areas be established when engineered techniques are used to comply with pollution loading limits, prohibit uses that are detrimental to water quality, restrict development in riparian zones, and establish development density limitations in the Lake Maumelle watershed. CAW also supported Pulaski County's development of a comprehensive Land Use Plan for the Lake Maumelle watershed in Pulaski County as a prerequisite to development of the zoning code. Through this collaboration with Pulaski County, CAW has effectively implemented a number of the key action items included in the Plan.

CAW addressed potential impacts from new development in the other two counties that make up the Lake Maumelle watershed by working with the state regulatory authority to prohibit the direct surface discharge of treated sewage throughout the entire watershed. This prohibition ensures that no new surface wastewater discharges will be allowed in the watershed and addresses a key action item identified in the Plan.

CAW also partners with various Arkansas state agencies, including the Arkansas Forestry Commission (AFC) and the Arkansas Game and Fish Commission (AGFC), for technical assistance in implementing a number of the land management action items included in the Plan. AGFC, AFC, and CAW have worked closely on forest management activities. The agencies provide technical assistance and labor for forest road stabilization, wildlife management activities, and monitoring and enforcement of recreation rules and regulations through enrollment of 18,900 acres of CAW-owned property into a Wildlife Management Area (WMA).

The Nature Conservancy has also been a partner in CAW's action items related to sound land management. CAW contracted with the Nature Conservancy to develop a fire management plan in order to address long-term planning for both wildfire and prescribed burns. The Nature Conservancy has also conducted prescribed burn activities on approximately 1,200 acres of CAW-owned property in order to reduce the risk of wildfire and disease, improve the health of the forest, and improve the water quality filtration abilities of CAW's forest lands.

To address the conservation land action item in the Plan, CAW has acquired 2,515 acres (3.9 square miles) since the Plan's adoption at a cost of $27.8 million. In addition to land acquisitions and enhancements, the utility has worked with private landowners to place an additional 295 acres (0.46 square miles) under conservation easements at a cost of $738,125. To fund these purchases and other watershed protection activities, CAW established a watershed protection fee in 2009 that generates nearly $1 million per year through a $0.45 monthly fee on an average residential customer (the fee increases based on meter size). The fee appears as a line item on customer water bills in order to provide greater transparency regarding the use of utility funds as well as to raise community awareness of watershed protection as a key component of the multibarrier approach to drinking water supply. CAW successfully leveraged the watershed protection fee funds to secure a $4 million Forest Legacy Grant from the US Forest Service, a $4 million state appropriation, and $1 million from the AGFC WMA lease for a number of these property purchases. The largest of the conservation acquisitions is the

915-acre (1.4-square mile) former Winrock Grass Farm along the Maumelle River, purchased in 2010.

Since purchasing the farm, CAW developed a master plan for restoration and improvement of water quality and potential use of the property as an outdoor education and learning center. CAW restored a 400-foot section of stream bank, which significantly reduced sediment and phosphorus inputs into Lake Maumelle. CAW is in the process of removing a low-water crossing downstream of the restoration that will further reduce sediment loadings, attenuate flooding, and enhance ecological and hydrological connectivity. CAW obtained $120,000 from the US Fish and Wildlife Service and $7,000 from the AGFC for these projects. As a condition to the grant that will allow CAW to acquire the Grass Farm property, CAW must also reforest a substantial portion of the property. CAW is planning to plant 44,000 seedlings on approximately 140 acres and will complete additional plantings over the next two years. Other planned restoration activities on the property include grassland and shrubland establishment as well as habitat enhancement for pollinators, migrating birds, and other native wildlife.

CAW created a Risk Mitigation Plan (RMP) for the Lake Maumelle watershed in 2009 and has performed multiple field and tabletop exercises using the RMP. Development and use of the RMP implements the hazardous spill risk reduction action item identified in the Plan. Based partially in response to lessons learned from the RMP exercises and partially as proactive measures for risk reduction, CAW installed facilities and purchased equipment to aid in and improve the utility's and community's response to a hazardous material spill in the watershed. In addition, CAW completed a comprehensive vulnerability assessment in late 2014 that includes a detailed risk assessment and risk mitigation evaluation for the ExxonMobil pipeline in the Lake Maumelle watershed. CAW continues to work toward funding and implementing risk reduction and mitigation measures identified in the vulnerability assessment, particularly those that directly affect the watershed.

Prior to adoption of the Plan, CAW had established rules and regulations to govern authorized recreational uses in the lake area. As part of the Good Lake Management Practices action item, CAW periodically evaluates the rules and regulations in order to determine if modifications are necessary. In 2015, CAW began development of a Recreation Management Plan that identifies, among other items, necessary modifications to the lake use rules and regulations in order to ensure water quality protection over time.

Evaluation and Revision

In addition to identifying clear action items and necessary mechanisms for its implementation, the Plan is part of CAW's adaptive management approach that includes constant evaluation and revision. The Plan states that "conditions in the lake and watershed must be monitored and assessed over time to ensure that the Plan is successful, as well as to show where adjustments to the Plan may be needed."

As discussed previously, CAW established a robust water quality monitoring program in partnership with the USGS and uses the collected data in its ongoing analysis within the watershed GIS databases. Continuous flow gauging and water quality monitoring over long periods of time are essential to evaluate watershed pollutant loading, as well as hydrologic changes. The USGS operates eight flow or reservoir elevation gauges in the watershed. Data collected by the USGS and CAW can be compared to the numerical targets of the water quality indicators identified in the Plan. This unique combination of resources allows CAW to track conditions in the watershed and lake and to both adjust treatment operations in the short-term and identify additional long-term action items needed to improve water quality throughout the watershed.

In 2014 and 2015, CAW began work to establish long-term biological monitoring of Lake Maumelle, Lake Winona, and the rivers, streams, and tributaries throughout their watersheds. The addition of biological monitoring provides a critical link between the robust water quality monitoring program and the health of these resources. Ultimately, biological indices will allow CAW to develop a more holistic approach for its water management efforts and will allow the prioritization of funds spent for protection and restoration.

Because many action items related to new development impacts were legislatively driven, CAW (working through Tetra Tech) used the watershed models developed as part of the Plan to evaluate the relative impacts of new development under scenarios and draft regulations outlined in the Plan as compared to regulations and scenarios that were politically and legislatively achievable. As a result, CAW's policy efforts were informed both by the potential for legislative success as well as the impact on water quality. If potential additional degradation were possible, CAW would be faced with the decision to increase the amount of conservation land committed or oppose a given proposal. As a result, the legislative process was an iterative process in which CAW was forced to balance the protection of water quality with the mitigation of impacts through the acquisition of conservation land. This process, by its nature, required constant evaluation and revision of the specific goals and targets identified in the Plan. In fact, because the adopted ordinances allow more exempt development than anticipated by the Plan, CAW chose to accept the policies with an understanding that additional conservation land will need to be purchased in order to offset exempt development.

CAW has identified a need to further refine the Plan with supporting resource plans. Each resource plan builds off of recommended action items in the Plan and provides additional details and specific, implementable steps to achieve the Plan's overarching goals and objectives. For example, CAW developed a Silviculture Plan for managing its forest resources in order to identify when and where ecological timber-thinning activities should occur. To complement this plan, CAW partnered with the Nature Conservancy to develop a Fire Management Plan that addresses long-term planning for both wildfire and prescribed burns. Both plans are part of CAW's Healthy Forest Initiative. As previously discussed, CAW is developing a Recreation Management Plan in order to identify and evaluate the water quality compatibility of recreational activities for Lake Maumelle and adjacent CAW property. Finally, CAW is developing a comprehensive Forest Management Plan that incorporates specific resource plans and identifies the future conditions of the CAW property and how to move from current status to the desired conditions while providing optimal protection of water quality.

Conclusions and Lesson Learned

CAW developed the Lake Maumelle WMP to safeguard the lake's source water quality. The Plan addresses the elements identified in AWWA/ANSI G300 Standard and has been an extremely useful tool in guiding CAW's leading-edge SWP activities.

Through development and implementation of its SWP Program, CAW has identified a number of key components necessary for an effective SWP Program. These include the following:

- The SWP Program was developed in collaboration with a stakeholder team that is representative of the community.
- The SWP area delineation satisfied state and federal requirements.
- The source water assessment went well beyond what is required under the Safe Drinking Water Act, and included a well-defined SWP area that contains an inventory of the location of potential sources of contamination (as presented in the Plan).

- The Plan includes a set of management practices that effectively control known and/or potential sources of contamination.
- CAW has an emergency plan in place to deal with accidents that may threaten the water supply in the SWP area.
- CAW has coordinated with state and local governments to develop and implement protections for the watershed from new development, including subdivision and zoning regulations as well as a prohibition of direct wastewater discharges into the SWP area.
- CAW has an understanding that effective SWP requires both active and adaptive management strategies.
- CAW recognizes the importance partnerships with governmental and non-governmental organizations in successful SWP management.

Lessons Learned

Some of the important lessons learned through CAW's SWP efforts include the following:

- Gaining strong community support is extremely valuable in ensuring the ability to thoroughly respond to threats to the water supply. This support can and should be leveraged to develop a broadly supported watershed protection plan.
- It is crucial to foster community education and involvement during the development and implementation of the SWP programs as resistance from various stakeholders (e.g., property rights advocates) may be costly, time consuming, and only address an isolated threat rather than a systemic issue.
  – This education and involvement must include both watershed residents and customers, when the two are differentiated.
  – Community education should also be repetitive and constant, and information should be consistent.
- Support from the governing board and senior management is critical to establish and fund a watershed management program, given the substantial amount of time and financial resources required.
- It takes a significant amount of time to bring a wide range of stakeholders together and build consensus on core problems and action plans.
- Sound science is a critical component of proactive watershed planning and monitoring. Generating sound science to inform decision-making for

SWP efforts requires the right partnerships, a commitment to time and patience, and sufficient financial and technical resources.

- Financial resources and qualified people are necessary to implement and adapt SWP action plans.
- It is critical for a water utility to have staff who have the ability to balance the water utility's needs with stakeholder expectations and sound management of resources.
- Expect the unexpected.

Successes for CAW's SWP Program include the following:

- Development and adoption of the Lake Maumelle WMP
- Legislative prohibition of direct surface discharge of wastewater in the Lake Maumelle watershed
- Implementation of a Watershed Protection Fee for customers of CAW that helps fund watershed land acquisition and improvement
- Passing of the subdivision ordinance and zoning code by Pulaski County with provisions for controlling development in the watershed
- Strong support by the Central Arkansas Water Board of Commissioners and local citizen groups for watershed protection
- Acquisition of more than 2,800 acres for conservation purposes
- Active management of approximately 1,500 acres of CAW property as part of the utility's Healthy Forest Initiative and ongoing restoration of the 915-acre former Winrock Grass Farm property
- Implementation of active land and water management and monitoring strategies for all CAW properties

References

Arkansas Department of Health. 2000. Source Water Assessment Summary Report. Little Rock Municipal Water Works.

American Water Works Association. 2007. ANSI/AWWA Management Standard G300 - Source Water Protection. Denver, Colo.: AWWA.

Benham Group. 1988. Report on Water Supply, Treatment, and Distribution for Little Rock Municipal Water Works, Little Rock, Arkansas. Tulsa, Okla.: The Benham Group.

Ozment, S., T. Gartner, H. Huber-Stearns, and N. Lichten. Forthcoming. 2016. Protecting Drinking Water at the Source: Lessons from United States

Watershed Investment Programs [working title]. WRI Report. Washington, D.C.: World Resources Institute (www.wri.org).

Stapleton, C.R. 1985. Lake Maumelle Diagnostic/Feasibility Study. Final Report. Prepared for Arkansas Department of Pollution Control and Ecology.

Tetra Tech, Inc. 2006a. Lake Maumelle Watershed and Lake Modeling—Model Calibration Report. Prepared for Central Arkansas Water.

———. 2006b. Lake Maumelle Water Quality Management Plan: Baseline Analysis Report. Prepared for Central Arkansas Water.

———. 2007. Lake Maumelle Watershed Management Plan. Prepared for Board of Commissioners, Central Arkansas Water.

The Cadmus Group, Inc. 2004. Lake Maumelle—Source Water Protection: Report to Central Arkansas Water. Prepared for Central Arkansas Water.

US Geological Survey (USGS). 1994. Water Quality Assessment for Lake Maumelle and Winona Reservoir Systems, Central Arkansas, may 1989-October 1992. Water-Resources Investigation Report 93-4218.

———. 2001. Analysis of Ambient Conditions and Simulation of Hydrodynamics, Constituent Transport, and Water-Quality Characteristics in Lake Maumelle, Arkansas, 1991–92. USGS Water-Resources Investigation Report 01-4045.

———.2004. Water Quality Assessment of Lake Maumelle and Winona, Arkansas, 1991 through 2003. Scientific Investigation Report 2004-5182.

Case Study: Remsen Municipal Utilities, Remsen, Iowa

Remsen Municipal Utilities, located in northwestern Iowa, provides drinking water to approximately 2,000 people. It uses a mix of water from two deep Dakota aquifer wells and five shallow alluvial municipal wells.

In 2007, the Remsen Municipal Utilities superintendent was concerned by the rising nitrate levels in the shallow alluvial wells and contacted the Iowa Department of Natural Resources (DNR) Source Water Protection Program coordinator to request planning assistance. At the same time, the Iowa DNR's SWP Program was initiating a new SWP pilot project program that would incorporate groundwater site investigations, provide local SWP team coordination and long-term local responsibility to conduct ongoing well monitoring, and secure nontraditional resources for implementation of the local SWP Program, as well as long-term evaluation of impacts of installed SWP practices on municipal well nitrate levels.

While facing the rising nitrate problem, the SWP project team assessed a treatment solution to reduce the increasing nitrate concentration (i.e., 27 mg/L in the primary shallow well with lower but rising nitrate levels in the remaining four shallow wells). This proposed treatment option would require installation of a reverse osmosis system at a cost of about $2 million (in 2007 dollars). In addition, it was estimated that the annual operation and maintenance cost would be approximately $50,000 per year. The current water operator would also need to upgrade his operator's license and certification. In light of the high costs associated with this option, the SWP team explored other alternatives for protecting the drinking water source. The team requested assistance from Iowa DNR's SWP Program to assist them in developing and implementing a SWP program that would follows the six elements of the SWP Program as outlined in AWWA/ANSI G300, Source Water Protection.

Vision and Stakeholders

At the request of the Remsen Municipal Utilities Board, the Iowa DNR SWP Program collaborated with stakeholders in Remsen by forming the above-referenced SWP team. The team began meeting in 2007 and developed a vision for the SWP plan, namely, "to develop a SWP plan that would include a 'boots-on-the-ground' technical assessment, involvement of local landowner, citizen, and conservation agency participation, implementation of nitrate reducing Best Management Practices (BMPs), and securing financial resources for BMP implementation, along with a long term monitoring program to evaluate progress in reducing nitrate levels in the drinking water supply." The Remsen SWP team was comprised of seven landowners, a US Department of Agriculture (USDA)–Natural Resource Conservation Service (NRCS) district conservationist, an Iowa State University Extension agriculture specialist, the local Resource Conservation and Development (RC&D) coordinator, a county sanitarian, the Iowa DNR SWP coordinator, a Plymouth County Pheasants Forever chapter representative, the Remsen Municipal Utilities superintendent and water operator, and various city officials and citizens.

Source Water Characterization

As part of the Iowa DNR SWP Program, a "boots-on-the-ground" comprehensive source water characterization was performed (i.e., groundwater site investigation) in order to provide scientifically sound data to the Remsen SWP team. The groundwater site investigation, which was completed in 2007, provided a wealth of data on the characterization of the source water (e.g., groundwater flow directions and water quality). Wells 1, 3, 5, 6, and 8 are shallow wells, ranging in depth from 32 to 35 feet, that tap into the shallow alluvial aquifer. Well 8 was associated with very high nitrate levels (i.e., 27 mg/L) that were much higher than nitrate levels associated with the other four wells (ranging from 1 mg/L to 15 mg/L). It should be noted that well 8 pumped the highest volume of water for the Remsen water supply. The groundwater site investigation also determined that the nitrate was coming from nonpoint sources.

The data generated by the groundwater site investigation were used in development of a groundwater flow model based on Visual MODFLOW, which is a groundwater flow modeling system developed by the US Geological Survey and fitted with a data visualization interface. Through simulation runs of the model, it was determined that well 8 intercepts much of the groundwater that flows from the eastern portion of the alluvial aquifer when the well is in use. This area, which is east of the Remsen well field, is also the recharge zone for the shallow municipal wells.

Because the recharge zone immediately east of the pumping wells was overloaded with the overapplication of fertilizers and feedlot manure, recharge to the shallow alluvial aquifer contained high levels of nitrate. Because of well 8's position and the effects that pumping the well had on groundwater flow in the shallow aquifer, much of the nitrate from the recharge zone east of the well field was captured by well 8. When well 8 was shut off in the MODFLOW model runs, nitrate concentrations were higher than expected for wells 1, 3, 5, and 6. In other words, well 8 acted as a "guardian well" for wells 1, 3, 5, and 6 by pulling in a larger portion of the nitrate-contaminated groundwater when it was in operation.

SWP Goals

The Remsen SWP goal was to decrease the nitrate level in well 8 from 27 mg/L to below the nitrate maximum contaminant level (MCL) of 10 mg/L over time. On the basis of the established vision and source water characterization process, the SWP team determined the following reasonable and meaningful objectives for Remsen's water supply:

- Use the Iowa DNR's 2007 groundwater site investigation results to prioritize protection of the area upgradient of well 8.

- Obtain permanent control of the land areas that were identified as primary contributors to the nitrate problem and convert these areas to native grasses.

- Through the use of local newsletters, magazines, and newspaper articles, inform and educate the city government, residents, and landowners within the community regarding the proposed SWP activities and how BMPs can aid in the protection of the water supply in a sustainable manner.

- Use an active planning approach for SWP, that is, identify and acquire resources such as cost-share programs available to landowners within the capture zone, obtain grants and 0% interest loans available to water systems to gain permanent control of land within the capture zone, and identify acres and rental rates within the capture zone that are eligible for the USDA Conservation Reserve Program (CRP). Through this process, it was more likely that landowners and the water system would be successful in implementing land acquisition and adopting BMPs.

- Have the Remsen Municipal Utilities superintendent work with the local RC&D coordinator, DNR SWP Program coordinator, local landowners, the Iowa State Drinking Water State Revolving Fund (DWSRF) program, Plymouth County Pheasants Forever chapter, Iowa Watershed Improvement Review Board (WIRB) members, Plymouth County USDA–NRCS, and the Remsen City Council to secure the state WIRB grant for purchase of priority acres, an DWSRF 0% interest loan to purchase additional acres for well field protection, funding from the CRP Wellhead Protection Program, philanthropic resources for seed and seedbed preparation, voluntary planting of the native grass seed in the capture zone, and long-term maintenance of the native grasses in the capture zone.

- Conduct long-term monthly water quality monitoring on wells 1, 3, 5, 6, and 8 for nitrate (along with precipitation) and compile the data from 2007 to present.

- Develop and make available the information associated with an effective SWP program to the general public and other interested parties.

SWP Action Plan

In 2008, the Remsen SWP team finalized their action plan, which stresses the use of stakeholder participation to solve the nitrate problem. The following organizations are part of the partnership association with the partnership-building process (note that many of these partners were part of the original SWP team):

- Seven landowners who were farming within the recharge zone participated on the SWP Team
- Remsen Municipal Utilities Board and superintendent
- Plymouth County USDA–NRCS
- Sioux Rivers RC&D Council
- Iowa DNR SWP Program for Targeted Communities Program
- Iowa DNR DWSRF Program
- Iowa DNR Private Lands biologist/ecologist
- State WIRB
- Plymouth County Soil and Water Conservation District
- Iowa State University–Plymouth County Extension
- Plymouth County Pheasants Forever chapter

Through meetings and collaborative effort, a final SWP plan phase 2 SWP plan was developed and approved by the Remsen City Council and the Iowa DNR.

Implementation of the SWP Plan

In 2009, the partners worked together to take action on implementing the SWP plan in order to convert a specific piece of agricultural cropland to native grasses and forbs in the recharge zone just east of the well field. The SWP team had identified the nitrate-contributing areas and focused their resources on developing a natural, economically feasible, and sustainable program to decrease the nitrogen load in the recharge zone. Native grasses with root depth an average of 7–8 feet were established. These deep-root grasses effectively take up excess nitrogen in the soil within the recharge zone and thus reduce nitrate load in the groundwater that flows toward the 32-foot-deep well 8 and the other four shallow public wells.

The SWP team's effort took time and resources. First, the partnership submitted a grant application to the Iowa WIRB and obtained $178,400 that was used to purchase 22.3 acres of cropland east of the well field. Second, the partnership assisted Remsen Municipal Utilities in applying for a 0% interest DWSRF loan for $548,559 to purchase an adjacent 35.34-acre parcel of crop land that had recently become available for sale on the west side of the well field and was deemed necessary for use in future protection of the well field. It should be noted that in order to secure a 0% interest DWSRF loan, an Iowa DNR–approved SWP plan is required. The purchased land includes a perpetual easement that will keep the land in native grasses and forbs. Third, the Plymouth County USDA–NRCS staff developed the plan, which was implemented using donated labor and funds ($15,000) from the Plymouth County Pheasants Forever chapter. The plan involved preparing the seedbed and planting the recharge zone with seeds of native grasses and forbs.

After conversion of the cropland to native grassland, the USDA–NRCS, an Iowa DNR biologist, and the Plymouth County Pheasants Forever chapter representatives developed a maintenance plan for native grass establishment (i.e., through controlled burning, mowing, and other vegetation-control strategies) to keep weeds in check, as it takes about 3 to 5 years for the native grasses and forbs to reach maturity. The Remsen Municipal Utilities Board and Remsen Fire Department took on the responsibility to implement this maintenance plan.

Evaluation and Revision

Remsen's SWP Program has undergone continuous evaluation through an ongoing monitoring program of nitrate levels in all shallow wells, with special focus on nitrate levels in well 8. Results show that from 2007 (i.e., the onset of the Remsen SWP effort) to early 2013, the nitrate level in well 8 dropped from 27 mg/L to as low as 11 mg/L (i.e., a 60% reduction). This drop in nitrate has contributed to a 2-mg/L decrease in the nitrate concentration in the blended finished water over the same time period, allowing Remsen Municipal Utilities to stay in compliance with the drinking water standard (i.e., MCL of 10 mg/L for nitrate).

From 2007 to 2012, the overall nitrate levels for Remsen were on a downward trend. This was largely attributed to the efforts of the SWP team and the leadership of the Remsen Municipal Utilities superintendent in implementing BMPs and possibly precipitation levels.

In 2013, seven years after establishment of the SWP team, there were retirements and turnovers associated with volunteers and personnel on the Remsen City Council, the Remsen Municipal Utilities Board, and the superintendent position. The volunteers and personnel held much of the local knowledge base and implementation efforts that were the foundation to a successful SWP program. In 2013 the Iowa DNR SWP Program coordinator met with the newly elected Remsen Municipal Utilities Board and Remsen City Council members to discuss SWP and provided them with Remsen's historical technical SWP assessment information. The newly elected utilities board and city council members were interested in learning about water quality. By 2015, it was apparent to them that their "back-burner" SWP Program was highly valuable and had saved a lot of resources to ensure that the community had high-quality drinking water. At the same time, there were renewed concerns regarding Remsen's elevated nitrate levels during 2013–2015. The monitoring results for the shallow municipal wells indicated a reverse trend in nitrate levels. The city reconvened the Remsen SWP team to discuss any recent changes in land-use within the municipal capture zone acres. In 2015, the Iowa DNR SWP Program was asked by the Remsen SWP team to assist in updating their SWP plan, with the intention of improving control of the source water nitrate levels.

In late 2015, the reconvened Remsen SWP planning team and new utilities board and city council members were developing updates to the Remsen SWP plan. The updates included the following:

- Continue monthly nitrate monitoring of the shallow groundwater public wells. An isotope sample of well 8 showed that the nitrate source was organic, indicating commercial fertilizer was not the source; most likely, the source was manure. An additional isotope sample from the municipal well will be taken for confirmation.

- Initiate surface water monitoring of the streams and feeder creeks in the recharge zone. The SWP team was adding monitoring of surface water that drains into the well field. Additional livestock numbers from 2012 to 2015 within the runoff area had caused concern that the increased numbers of livestock in the area over recent years may be affecting the increases in nitrate levels in the municipal wells. Once there was sufficient information on ammonia and nitrate levels in the water draining into the well field, an informed decision and updates to the Remsen SWP plan would be made.

- Coordinate with the local Deep Creek Watershed Water Quality Initiative project coordinator, the USDA–NRCS district conservationist, and DNR Private Lands biologist to access cost-share programs for BMP implementation (e.g., planting cover crops, developing buffers, conducting nutrient management, installing a livestock total containment structure).

- Conduct quarterly SWP team meetings (e.g., review data, update the SWP plan, identify resources, inform landowners of water sample results, encourage landowners to adopt practices to decrease nitrates, discuss storm water plans).

- Educate the city council and area landowners on the importance of protecting the Remsen capture zone area from pesticide application within the priority acres (i.e., historically atrazine and 2, 4-D were applied).

Conclusion

The Remsen SWP project was successful in demonstrating that through land-use changes, it is possible to decrease the nitrogen concentration in shallow public wells. However, for reasons that are not yet clear to the project team, some increase was seen and it is currently being reevaluated.

A SWP plan is a living document, and a SWP program should follow the iterative process that consists of the six elements of a successful SWP program. The Remsen SWP project has proven that the following are essential to success: local SWP teams; a SWP plan that keeps existing and new SWP members informed; a SWP plan that is never "done," as changes are always occurring within the capture zones; and a better understanding of the hydrology and hydrogeology of the aquifers involved. It is therefore necessary to continue monitoring water quality, evaluate options, and implement necessary practices through updates to the SWP plan over time due to new knowledge and changing conditions. The Remsen SWP project exemplifies why SWP is an ongoing process. Land-use changes, local government changes, and water quality changes all have impacts on the sustainability and viability of a community's precious water supply. These impacts can be negative if the community is not diligent about its SWP Program.

This case study also shows that a community water supplier is wise to enlist the assistance of a knowledgeable SWP team to conduct monthly water monitoring, evaluate water quality monitoring results, assess land-use changes, facilitate SWP team and stakeholders meetings, collaborate with conservation partners, and prepare SWP plan updates, among other SWP activities.

As noted previously, local government and SWP team members may change over time. Such personnel turnover can be challenging for long-term SWP programs, especially where a particular "champion" has been leading the efforts. In addition, a SWP program should never be stagnant. Ongoing meetings, even those that occur on a quarterly basis, are essential for the continuous and consistent deployment of a SWP program that protects highly susceptible municipal wells. Protecting sensitive land through land acquisition and easement purchase proved effective for Remsen. However, often there are not sufficient resources to obtain control of the majority of the capture zone acres, making outreach and education essential to protection of susceptible public wells. For areas that are not protected under a local SWP program, land use can change in a short period of time. Therefore, the local SWP team members must be vigilant and stay alert regarding any changes to land management in the SWP areas and of runoff from surrounding areas to the recharge zone of any susceptible municipal well. Finally, SWP programs must address climate variations. Large increases in precipitation and the resulting increase in runoff from surrounding lands can increase the nutrient loads to a susceptible well or well fields. Land-use changes and increases in livestock numbers within susceptible capture zones, among other variables, can contribute to higher nitrate levels in a few short years.

It is notable that approval of the 2008 Remsen SWP action plan by the Iowa DNR was largely achieved within about a 2- to -3-year year time frame. The DNR's initial groundwater site assessment helped in the Remsen SWP team's collaboration and planning, as well as provided project credibility to local, state, and federal partners from the initial SWP planning to the final conversion of approximately 100 priority row-crop acres to native grasses within the well field capture zone.

Because of the excellent SWP planning and implementation efforts from 2007 to 2010, Remsen Municipal Utilities was nominated by the Iowa SWP Program and received the 2010 AWWA Exemplary Source Water Protection Award for small water systems. In addition, for their SWP efforts from 2007 to 2012, the Remsen SWP team received the 2013 Iowa Governor's Water Quality Excellence Award. Furthermore, the Remsen SWP team in 2010 proposed forming three subcommittees to develop plans to protect the well field and generate educational opportunities to foster an outdoor classroom, nature trails, and a long-term system for native grassland management.

In summary, as of early 2013, the conversion of cropland to native grasses has improved source water quality. Moreover, this conversion process has restored habitat for pheasants, offered outdoor educational opportunities and outdoor exercise through walking trails, and provided wildlife habitat, all of which have improved the quality of life in a small community in northwest Iowa.